Making Math Matter

A Math Resource/Methods Book
for experienced and beginning teachers

by

John Mudore

Printed in the United States of America — 1st edition.
Library of Congress Catalog Card Number: 93-77825
ISBN 0-9636514-8-X

INFINITY PUBLISHERS
P.O. Box 333
Black Earth, WI 53515

ACKNOWLEDGMENTS

Constance Faye Mudore This book would not have been written without the support of my wife. Her input and editing are reflected on every page. Words cannot express my gratitude.

Professor Paul Shoecraft He showed me how to make math matter and encouraged me to keep writing after reviewing the first few chapters.

Mark Parish A computer consultant and English teacher who helped me refine my philosophy of teaching.

James Ciha A colleague and mathematical proofreader.

Nancy Nelson Her editing improved the readability of part one.

Lindy Peckham A readability proofreader.

My students The material in this book was created with my students who were willing to explore how math relates to life. It was fun!

DEDICATION

TO MY PARENTS
JOHN AND DORIS MUDORE
for showing me how to
make life matter

TABLE OF CONTENTS

INTRODUCTION: WHO IS THIS BOOK FOR?

Making Math Matter is a resource/methods book written by a veteran classroom teacher for middle and high school math teachers, college and university instructors of math courses below calculus, and professors and students of math methods courses. Both novice and experienced teachers will find the information useful because it illustrates how to make teaching and learning math meaningful and fun. Part one contains classroom management techniques designed specifically to make teaching and learning math more enjoyable. Part two contains teaching strategies on how teachers can make math matter by presenting algebraic, geometric, and trigonometric topics in a real world setting which emphasizes learning and seeing value in that learning.

The classroom management techniques, combined with activities relating math to life, work together to engage students in learning. Management strategies include an evaluation system which emphasizes learning rather than grades, a positive tone in the classroom that is non-threatening, and class session pace which maximizes student involvement. The book shows teachers how to demystify math and explains why learning math is important to everyone.

Specific teaching strategies illustrate how teachers can make math matter. Concepts are introduced by relating math to life and by solving real problems, i.e., problems whose solutions are not known prior to the solving. Individual and cooperative learning activities that require students to reason and to communicate mathematically are integral components of concept development. (For example, working in groups, students discover why the bottom step in a stairway is sometimes a different height from the other steps. Students work in small groups because designing stairs to code is complex and requires input from others in order to solve it.)

Concepts and skills are tied together to show how they are interrelated by comparing and contrasting them in words. Students come to truly understand the relationship among expressions, equations, and functions when they construct a chart that compares and contrasts them in English. In addition, information is presented by using variables that are the first letters of words associated with what they represent, e.g., in the equation $\log_B A = E$, A stands for Amount, B represents Base, and E symbolizes Exponent.

Students learn that math is a language which they can use to make sense of the world. They learn to use mathematical modeling to identify and to describe patterns that exist in life. For example, students create a formula which describes the relationship between their age and grade level. They learn arithmetic progressions by writing an equation which generates a cat's age into human years.

The interdisciplinary material presented in this book offers a meaningful way to introduce specific concepts that are taught in mathematics textbooks and can be easily integrated into any curriculum. All of the material discussed has been classroom tested with many of the major textbooks in a variety of settings (junior high, high school, and college levels; traditional and alternative programs; private and public schools). The methods work well with all types of students (tech prep, college bound, slow learner, gifted and talented, learning disabled, and special needs students) because the methods involve several modes of learning.

I recommend that you read the entire book first to get an overview of the classroom management techniques and math activities included in it. Then go back and decide which management strategies and math activities you want to use. Keep the book handy so you can refer to it throughout the year.

For several years, I have shared my expertise in teacher-training workshops. Teachers who have attended my workshops have asked me for written material on my management and teaching strategies. *Making Math Matter* is my response to these requests. It is my hope that the ideas presented here will stimulate creative teaching of mathematics.

Part One

Joy
of
Teaching

Chapter 1

Setting the Tone

Teachers inspire learning. What they do and say matters. Therefore, teachers need to be aware of their own non-verbal and verbal behaviors. Words can be interpreted many ways depending on how they are spoken and other non-verbal language that accompanies them. Tone of voice, facial expressions, and other gestures can contradict the words spoken. The "real message" is communicated by a person's non-verbal behavior.

Even though actions tend to speak louder than words, the specific words spoken are very important. The words people choose affect and reflect how they view themselves and their world. It is one thing for people to view what they do in life as something they "have to" or "should" do. It is quite another perspective for people to view what they do in life as something they want to do to fulfill some personal *need.*

Students often ask, "Do I have to do this?" or "Why should I do that?". Teachers can simply respond by stating that no one has to do anything. Students choose their actions and pay the consequences. Life is what they make it. If they *want* to learn, then they *need* to do their work. Nonetheless, teachers are responsible for persuading students that learning math is worthwhile. (See Chapter 7 on "Why Learn Math?".)

What transpires during the first few days of class often sets the tone for the entire semester. The words and actions teachers choose greatly affect the classroom environment. While students are ultimately responsible for their own behaviors, teachers can establish and maintain a safe learning environment that is positive, non-threatening, and accepting. This is no easy task, but it can be accomplished by creating and enforcing classroom expectations that respect students as young adults.

Classroom guidelines need to be established and discussed with students during the first week, but not on the first day of school. Since most teachers traditionally discuss classroom rules during the first class, middle and high school students may feel that their instructors lack respect for them. After all, students have been abiding by classroom rules for many years. Teachers who start the first class session with classroom rules may be implying to their students that they do not know how to behave appropriately. However, students, the first day of school, generally know how to behave and tend to be very attentive.

It sets a much more positive tone to use part of the first class for the teacher and students to get acquainted. Students could write the following information on an index card: their name, home phone, name of parent(s) they wish the teacher to talk with about their work, their previous math courses, and their interests. On the back of the index card students can reveal talents and accomplishments that they feel comfortable sharing with the class.

The students can then be paired and given time to share their interests, talents, and achievements with their partners. After a few minutes, students introduce their partners to the class using the information written on the index cards.

Completion of an index card is an old technique, but it shows students that teachers are interested in them as persons. The cards provide information that teachers can talk about as they get to know the students throughout the year. The specific information requested on the card is not that important as long as it concerns the students. They especially appreciate being asked who teachers may contact at home.

Just as teachers are interested in getting to know students, students are eager to learn about their teachers. The first class session is a good time for instructors to hand out a brief resume including information that *they* feel comfortable sharing. (See sample resume at end of chapter.) Teachers distribute their resumes after the index cards are completed and discussed. Students read the resume, ask questions, and make comments.

The course syllabus could be printed on the back side of the resume. The syllabus needs to include the grading procedure and to define cheating and its consequences. Students and parents have a right to know this information. (See the chapter entitled "What is Cheating?".) The course syllabus and teacher's resume can also be distributed to parents at open houses and parent-teacher conferences.

Teachers can begin the second class meeting by telling their students that every one of them can learn math. There will be times when they do not understand every concept and skill to their fullest, but that time will pass and they will not be scarred for life. There are many people who are happily living who did not fully understand all the math they were taught. What they did learn from their math classes, though, was to think analytically.

Students need to hear that it is natural to struggle with math problems. In order to ensure that all students feel comfortable struggling with problems in class, the teacher and students together can establish behavioral expectations or guidelines that will protect their right to learn in a warm and caring environment.

It is known that boys are called on more than girls to answer questions. Teachers need to constantly keep this in mind. A little notebook with two columns headed girls and boys could be used to tally students' responses. Girls can learn math as well as boys, but they need to be called on as often as boys, and they need to be given sufficient time to respond. Teachers can ask students for help in applying this procedure and formally announce it as a classroom expectation.

It is wise to have only a few classroom expectations since many guidelines are difficult to remember and enforce. Furthermore, too many guidelines show disrespect for students. Whatever the expectations are, it is paramount that teachers be consistent in enforcing them. Students readily agree that the following two guidelines are necessary to protect their right to learn.

1. One person talks at a time during class discussions and presentations.

It is difficult to hear a person speak when others are talking. Besides, everyone wants to be listened to when speaking. Students talk while others are speaking partly because they are expected to listen to others much of the school day. Teachers need to keep this in mind when they design the structure of their class sessions. Less time could be spent on whole class presentations and discussions and more time allowed for individual and small group work. When students are inattentive, the teacher could stand by them or call on them to regain their attention.

2. No one makes fun of others.

Joking and humor are essential to creating a relaxed classroom environment. However, there is a fine line between joking and ridicule. Joking with people at their expense is ridicule. If a student ridicules another person (student or teacher), teachers need to interrupt the ridiculer, and immediately return to the matter at hand. Teachers then talk with all students involved in the incident after class/school.

Teachers are human and may, inadvertently, make fun of students. Instructors can encourage students to inform them when students feel that they have been ridiculed. Since students are reluctant to initiate such a conversation with teachers, teachers can apologize to students when they feel that they have crossed the line. Try doing this sometime. The student will be shocked!

Some students do not realize that name calling is ridicule. Name calling labels people and removes their uniqueness. In order to help students *see* their uniqueness, the teacher could have students bring in a picture of themselves, (or the teacher could take a picture of students individually or in small groups) and make a collage of the pictures on a bulletin board. The collage could be entitled, "Label Jars, Not People."

If teachers have only a few guidelines based on mutual respect and they enforce them, there is no need for a list of consequences. Enforcement means to solve the problem — not to punish students. Granted, teachers need to have a set of measures for resolving problems with students who continue to be disruptive, as is discussed in the chapter entitled "Strategies for Dealing with Troubled Students." Most students, however, respond positively if they are treated with human dignity.

After discussing the classroom guidelines, the second class session could continue as described in the next chapter.

Sample Resume: John Mudore

Nineteen Years Teaching Experience.
Math Teaching Assistant , Arizona State University, Tempe, AZ.
Math Instructor, Arizona State Penitentiary, Florence, AZ.
Founder and Director of Alternative High School Program, Portland, OR.
Math Instructor, Highland Junior College, Freeport, IL.
Teacher of Gifted and Talented Students Grades 3-6 for seven years.
High School Math Teacher for the past twelve years.
Instructor of a college Math Methods Course.

Professional Activities
Author of *Making Math Matter*, a resource book for teachers.
Presenter of Teacher Workshops for middle and secondary math teachers.

Interests
Hiking, writing, jogging, human rights, cross-country skiing, basketball.
Coordinator of an Amnesty International Group for six years.

Chapter 2

Pacing a Class Session

Learning occurs only when students are engaged. The pace of a class must vary in order to gain and retain students' involvement. One method for varying the pace is to divide the class period into segments. Each segment has its own pace and promotes a different level of student involvement.

The first segment begins when students arrive in the classroom. As they enter, they write the problem number(s) of the homework question(s) that they do not understand in a designated space on the chalkboard. If more than one student has a question about the same problem, they place a check mark adjacent the number. This informs the teacher how many students are having difficulty understanding a particular concept. The teacher can also write problem numbers on the board that s/he wants worked. When the bell rings, the teacher officially begins the class by asking a student to give an example of where math is happening in the real world.

One student each session is called on to briefly share with the class a math related newspaper or magazine article. Students hand in their articles to the teacher at least a day before they are called on. The teacher talks with students during segment two or three about how they are going to highlight their article. Only one article per student each quarter is required.

After this brief introduction, the second segment begins with the teacher asking students to look at their homework to see if they know how to solve any of the problem numbers written on the board. In alphabetical order, students are called on to quickly choose a problem and write its solution on the board. The number of students that work different problems at the board simultaneously depends upon the amount of available board space. If a student does not know how to solve any of the problems, s/he can pass. In the next class session, the students who passed are allowed to choose first, since the easiest problems seem to be selected first. The teacher then resumes calling on students in alphabetical order.

A record keeping system insures that all students participate in board work. Credit is given to promote participation. Most students, however, are willing and eager to work problems on the board. They like to get out of their seats and demonstrate their expertise, provided the classroom environment is positive, non-threatening, and accepting.

While some students are at the board, the students at their seats answer each other's questions in preassigned groups. The atmosphere is lively yet relaxed. This segment is characteristically somewhat noisy, since the students are verbally and physically involved in learning. As the pace varies, so does the noise level.

In the third segment, the students who worked at the board remain there to answer questions concerning their solutions. Students who have questions are allowed to go to the board and get their questions answered one on one. Students can ask the teacher for a further explanation after they talk with the student who solved the problem. Only if many students are having difficulty with a problem does the teacher need to give an explanation to the entire class. Too often students are bored listening to what they already understand. This procedure allows for most questions to be answered individually either in small groups or at the board, one on one.

During the second and third segments, roll is taken, and students with special concerns are dealt with individually. The teacher moves around the room and only answers questions if a student cannot. This encourages students to communicate with each other which, in turn, fosters self-confidence and a deep understanding of mathematics. By the end of the third segment, students are ready to sit, listen, and participate in the presentation of the new concept, the fourth segment.

Teachers can do more than present information. They can stage it. In theater, directors stage scenes in order to bring to life a particular situation. In the classroom, math teachers can stage information in order to bring to life a specific concept or skill. Information is staged *by presenting the right amount of information at one time, by associating it with life, and by comparing and contrasting information.*

Too much information regarding how a concept works can be presented at once. Experience will reveal the amount of information that a class can process in one period. Each class has its own rate of learning. The teacher can ask students if too much information is being presented at one time.

When a new concept is introduced, it is important to show students only how the concept works, not why it works. "Why" the concept works can be explained after students are comfortable with "how". Most students find it too difficult to learn both aspects simultaneously. It is simply too much information to process at one time.

Concepts are introduced by relating math to life and by using real problems, i.e., problems whose solutions are not known prior to the solving. Concepts and skills are tied together to show how they are interrelated. Information is presented by using variables that are the first letters of words associated with what they represent. Whenever possible, an individual or cooperative learning activity is provided to help students discover the concept.

During the fifth segment, students remain seated, work problems, and discuss answers with those seated near them. The latter is essential, since some students need clarification from peers in addition to the teacher's explanation. Besides, students learn as they clarify. In order to keep students on task, the teacher may want to assign one problem at a time for students to solve. While students solve problems and help each other, the teacher can walk around the room and answer questions. After a few minutes, a student or the teacher writes the problem's solution on the board.

Closure can be brief to long, depending on the amount of class time remaining. The teacher can close with a reminder of the next day's assignment, a motivational story, a math puzzler, or math-related career information.

The order and length of these segments are not sacred. The teacher may want to present the new concept first, especially if a class is difficult to manage. Cooperative learning activities may take the entire period. The teacher needs to be organized, yet flexible and spontaneous. What is important is that the teacher varies the pace of a class period to gain and retain students' involvement.

This model for varying the pace of a class session is the result of many discussions with students. Students learn more, given ample opportunity to evaluate class structure.

Chapter 3

Interior Designing

Assigned seats, arrangement of desks, and placement of chalkboards are three components of the physical setting that are frequently overlooked which have a significant impact upon students' learning. Teachers can look at their classroom to see how they can design the physical setting to engage students in learning. Learning is what happens when students are engaged. It is an active, not a passive process. Learning is not done to students, it is done by them.

Assigned seating can be used to give students the chance to be teachers as well as learners. The "official" teacher can answer only one question at a time but so can each student. Encouraging students to help one another promotes a sense of learning as a shared effort. This can be done in small cooperative groups or one on one. For example, higher ability level students can be interspersed among lower ability students. This arrangement enables students to answer questions and to share information with each other. Students with lower grade averages often raise their averages when the seating arrangement is adjusted to maintain a mix of students.

Desks can be arranged so that student visibility of all chalkboards is maximized. Maximizing visibility of chalkboards can make it feasible for students to work problems at the board. Boardwork engages students in learning and gives them a positive means of channeling their energy. In a room with two blackboard walls, as many as eight students can solve as many different problems in a few minutes and explain their solutions. This is significant since many students need to see as well as to hear an explanation.

Maximized visibility of chalkboards also enables teachers to display all the information simultaneously when they present a lesson. Students can *see* the connection of the whole presentation from beginning to end.

If a classroom has chalkboards on two opposing walls, the desks could face the two walls without boards. Half of the desks face one of the boardless walls, and half face the other boardless wall. This insures that all students can see both boards. See diagram below.

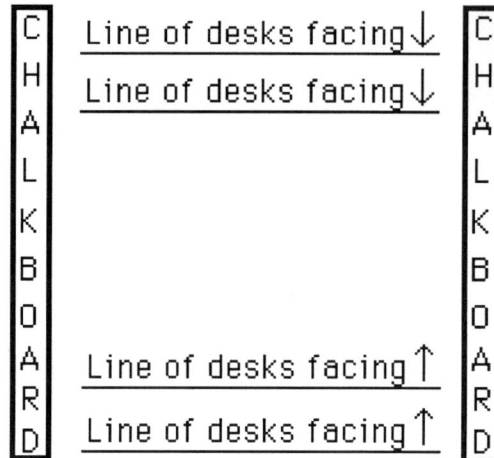

If a room has chalkboards on adjoining walls, the desks face either of the walls containing a board. With this arrangement, students need to turn only their bodies, not their desks, to see both boards. See diagram below.

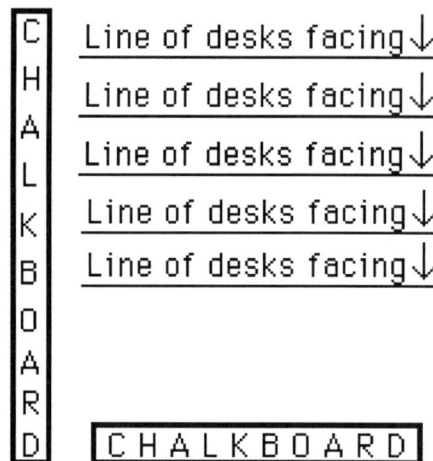

If a teacher is fortunate enough to have a room with chalkboards on three walls, the desks face away from the boardless wall. Like the preceding arrangement, students need to turn only their bodies to see any of the boards. See diagram below.

```
┌─┐ Line of desks facing↓  ┌─┐
│C│ ─────────────────────  │C│
│H│ Line of desks facing↓  │H│
│A│ ─────────────────────  │A│
│L│ Line of desks facing↓  │L│
│K│ ─────────────────────  │K│
│B│ Line of desks facing↓  │B│
│O│ ─────────────────────  │O│
│A│ Line of desks facing↓  │A│
│R│                        │R│
│D│   ┌─────────────────┐  │D│
│ │   │C H A L K B O A R D│  │ │
└─┘   └─────────────────┘  └─┘
```

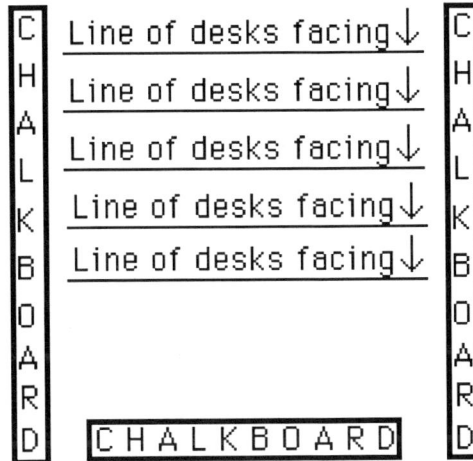

It is only reasonable to have students write solutions on the board if there is enough board space for approximately one-fourth to one-third of the class to simultaneously work problems. Since many classrooms lack sufficient board space, teachers can demand more chalkboards. Chalkboards are very inexpensive and can be easily attached to any wall.

If it is impractical to have students work problems at the board, more time can be allotted for students to help each other in small groups. The teacher moves from group to group answering questions that students cannot. The teacher works on the board and discusses with the entire class only problems that are common to several groups.

It takes time to group and regroup students so that each group has a mixture of ability levels and the members are willing to work with one another. However, it is time well spent because it helps students realize that they are ultimately responsible for their learning.

Chapter 4

Grades Evaluate What?

Often students regard their grades as measures of personal worth. They think that the grades they receive evaluate them as persons. Students need to hear from teachers that grades do not reflect their personal worth. "A" students are not necessarily "A" persons, nor are "F" students "F" persons.

Even though grades do not evaluate students as persons, grades influence students' futures in college and work. Consequently, teachers need to be "up front" about their grading systems and testing procedures. These need to be communicated to students during the first week of school.

Homework, individual and group projects, classroom participation, effort, journals, and tests can all be part of the grading system. Each component, however, does not have to be quantified into a formula in order to be part of the final grade. For instance, if students' test scores increased over a grading period, they completed all or most of their assignments, and they participated in class work, teachers can justify giving borderline students a higher grade (e.g., C+ becomes a B-).

Perhaps the most important component of any grading system is that students are placed in the proper math courses. If students do not possess the appropriate background necessary to assimilate the material, their chances of succeeding are slim. Students whose grade in a prerequisite math course is a D or lower can be encouraged to retake that course. Transfer students who are experiencing difficulty especially appreciate this option. Teachers can check records and talk to counselors of students who do poorly on the first and/or second tests.

Grading systems must be designed so that if students take notes, participate in class work, and do their assignments, they will earn an A, B, or C. This can be accomplished by teaching material and posing test questions with a degree of difficulty appropriate for the students' ability level. Math teachers have control over the level of difficulty of problems assigned and questions placed on exams. They also control when tests are given. Students will experience more success if they have input into when exams are given. Teachers can ask students if they are ready to take an exam.

Authentic and performance assessment of learning measures critical thinking and higher order reasoning skills and does not ask for information that can be looked up. Memorization of formulas is not the goal of learning. Learning involves knowing what formula to use, how to use it, and where to find it if forgotten. This is what teachers are expected to know. So why are expectations of students different? If students forget a formula during an exam, they could ask their teacher for it and lose few (if any) points. Students must exhibit that they know what formula to use by specifying the name of the formula. This procedure allows students to show that they know how to use a formula and not lose full credit for a problem just because they forgot information that could be looked up.

If tests are to measure performance of thinking skills and not memorization of information, time must not be a factor. Some students require more time than others to complete exams. Exams are written on two or more sheets so that students can take parts of an exam at different times. Students make arrangements ahead of time if they want to take more than one class period to complete a test.

Some students perform better if there is plenty of room on the test sheets for them to show their work. Once students complete their exams, they can ask the teacher for scratch paper to check their work. Teachers can insist that tests not be turned in until the end of the class period to encourage students to check their work.

Tests are a means, not an end, to learning if students are required to make test corrections. Mistakes are lessons to be learned. Each lesson provides an opportunity to acquire knowledge. In order to accomplish this, the teacher returns corrected tests to students as soon as possible so they can correct their errors. The teacher may want to highlight common mistakes with the entire class. A grade is placed on exams after students have completed their corrections. This is done for two reasons: students are asked to verify that their total number of points is correctly computed, and they are encouraged to make sure the proper amount of points were taken off for each incorrect answer. Students can earn more points if they can rightly justify why they deserve more credit. When students complete their corrections, they turn in both their tests and corrections which are kept in their personal file until the end of the semester. The teacher allows students, who need more time to complete their corrections and who were absent the day of test corrections, to revise their errors during the homework segment of future class sessions. Students must do test corrections before taking the next exam.

Retake tests are required for students who received D's or F's. Students can increase an F to a D- by achieving at least a D- on a retake; and they can increase a D to a C- by scoring at least a C- on a retake. The requirement to retake exams means that it is more work to fail than to succeed. Students earning D's or F's on tests will need to see their math teacher three different times during non-class time to do test corrections, retake the test, and do corrections on the retake. This requires much more time for students than if they had done their work and earned an A, B, or C on the original test. Most students figure this out and no longer need to retake tests.

Retake tests are required to ensure students continued success since much of math is incremental. Students are destined to fail unless they are given more than one opportunity to succeed. Requiring retakes also communicates to students that the teacher is primarily concerned with learning, not grades. After all, it is much easier and less work to allow students to fail than to arrange and correct retakes. While retakes involve additional time for both students and teachers, retakes are worth the effort. Retakes give students a second chance to succeed, which reduces their test-taking anxiety levels and builds their self-confidence.

Those few who refuse to do their work can only fail if they choose to fail because the grading procedure is designed for success. If students do not have their test corrections and retake tests completed in time for the next test, they do them while the others are taking the scheduled test. Since only semester grades go on students' permanent records, students cannot fail math unless they have chosen not to complete their work by the end of the semester. This grading procedure makes it clear to students that they are choosing to fail rather than being failed.

Time need not be a factor in learning math. Students will learn math if they are given the time they need to complete their tests, if they can hand in homework late as long as it is before they take the test covering the current homework material, and if they can take tests and retakes any time before the semester ends. Teachers are not fired(failed) if they do not complete tasks on time. They are asked to get them finished as soon as possible. Even the IRS gives extensions to file and pay taxes! The IRS wants money and will take it late. If teachers want students to learn, then instructors need to accept late work and not fail students for not completing it on time.

It is still important to expect work to be completed by a certain date, but teachers do not need to get upset and disappointed with students who do not comply. Teachers can have students put in writing what and why they did not finish on time and then place this in their folder. This way, there is a record of students' excuses for incomplete work in their own handwriting. Accept all excuses! But expect the work to be completed soon.

Society demands that learning be scored and that a grade be assigned for each score. Learning occurs, nonetheless, even when it is not graded. In fact, grades often impede learning in that students do only what is necessary to receive the grade they want. They pursue a grade, not learning. But learning, not grades, can be the focus of all math courses. Grades based on improvement, assessment of thinking skills, test corrections, retakes, and journals are techniques that emphasize learning. (See the chapter entitled "Demystifying Math.")

Teaching math so that learning is the focus requires that the teacher be organized. It has been previously mentioned that students' materials are placed in their personal files. A personal file is a separate folder for each student's tests, projects, excuses, and other pertinent information. These materials are kept on file until semester grades are distributed.

Ultimately, teachers do not give grades, students earn their scores. In order to help students understand that their grades are the result of their own efforts, teachers can have students record their scores in their journals. At the end of the grading period, students can compute their own grades. Teachers need to remind students, however, that grades are not a measure of personal worth. Students need to hear this every time report cards are distributed.

A healthy self-esteem is the most significant quality that students acquire. Students with poor self images do not accomplish as much as they would if they felt better about themselves. Students' self-images and confidence levels are related to how they view their math-learning abilities. It is paramount to the future success of students that they believe that everyone can learn math, no matter what grade they earn. Once students finish their formal schooling and get their first jobs, grades become insignificant. What matters is their self-esteem, their personal characteristics, and their abilities to think, solve problems, and learn.

Chapter 5

What is Cheating?

Teachers need to discuss cheating with their students since the definition of cheating varies from teacher to teacher. In order to facilitate this discussion, a value worksheet, similar to the example at the end of this chapter, can be completed anonymously by each student. Students complete Side One before looking at and completing Side Two. The teacher informs the class that all responses will be read to the class and reminds them not to write their names on the worksheet. (Please read Side One and Side Two of the worksheet at the end of this chapter.)

Once the worksheets are completed, a student collects and shuffles them. The teacher then reads aloud the responses to question one and allows time for comments. This procedure is repeated for Questions Two through Four. Students often suggest, in response to question four, that desks be physically rearranged during tests so that students cannot see each others' papers.

After discussion of Questions One through Four, a student tallies the responses for Questions Five and Six on the board. Those who answer Question Five supposedly do not approve of cheating on tests and usually answer "yes." This response states that they would be comfortable sharing their beliefs about cheating on tests with younger relatives. Those who answer Question Six supposedly believe that cheating on tests is O.K., yet they typically answer "no" which indicates that they would *not* feel comfortable sharing their beliefs about cheating with a younger relative.

The teacher asks the class why they think that students who say they would cheat, would not encourage their younger relatives to do so. Usually, some students admit that they do not approve of cheating, even though they said that they would cheat. The class recognizes that it is very challenging for people to act in accordance with their values. The teacher can ask those who might consider cheating, to stop and ask themselves on each occasion when they have the opportunity to cheat, whether they actually need to do so. Cheating is like any habit; students do it without thinking. If students choose not to cheat once, it makes it easier for them not to cheat the next time, and they will feel much better about themselves.

Besides, does cheating really pay off? It does not if you believe that people are treated as they treat others or "what goes around comes around." Years ago when I was a math teaching assistant at Arizona State University, I sold my 1966 Mustang which was in bad shape. When people came to look at the car, I told them all the things that were wrong with it. They were totally amazed by my honesty. I told them that I would accept whatever they offered; $100 would be fine.

The first few people who looked at the car didn't want it for any amount because they didn't want a "fixer upper." The third person was an accountant who was looking for a car that his sixteen year old son could fix. After I explained everything that was wrong with the car, he asked how much I wanted. I told him whatever he thought was a fair price; $100 would be fine.

He said that he had never met such an honest person. He handed me $300. I said that it was too much. He said that the extra money was for being honest.

Some students might believe that cheating does pay. It is important for the teacher to acknowledge that they are free to believe whatever they want. However, their beliefs do not exempt them from paying consequences. For example, the consequence for cheating on a test could range from a score of "0" to an F for the grading period depending upon the severity of cheating, e.g., copying solutions, stealing an exam, etc. The teacher can show the class how a "0" on an exam effects students' averages for a grading period.

The most powerful thing a teacher can do to inhibit cheating is to discuss it openly and to create an environment wherein students do not feel it is necessary to cheat in order to succeed. For example, students could be given more than one chance on tests and they could be encouraged to work with each other on homework and projects. Cheating could be defined as not doing your own work only on tests.

Value Worksheet

People sometimes act impulsively without thinking about their values or whether their actions are consistent with their values. This worksheet will help you explore what you think about honesty on tests. Do not write your name on this form. All responses will be read to the class. Circle responses.

1. Are there any circumstances under which you think it would be appropriate for you to accept answers from someone else during a test?

   a. Illness or _____ kept me from studying, but I could have passed the test if I had studied.
   b. I do not have the ability to pass the test, but I need the credit.
   c. A friend offers to make the answers easily visible during the test.
   d. The test is unfair. Explain.

   e. Under no circumstances is it appropriate to accept answers.

2. What do you think about helping someone else during a test?

   a. O.K. to make my answers visible to them during a test.
   b. Would do anything to help a friend as long as I would not get caught.
   c. Would do anything to help a friend even if I got caught.
   d. Would depend on why the friend needed help. Give reason.

   e. Would not do it under any circumstances.

3. How would you feel toward someone who asked for help or copied answers from your paper during a test?

   a. Would not bother me.
   b. Pleased that they thought I was smart enough to be of help.
   c. Other. (Elaborate.)

4. Are there any means you would recommend to curb cheating on tests?

Side One

Answer question 5 if your answer to number 1 <u>and</u> 2 was e.

5. Would you feel comfortable sharing your beliefs about cheating with your younger sister or brother, cousin, nephew, or niece?   Yes or No?

Answer question 6 if your answer to number 1 <u>or</u> 2 was a, b, c, and/or d.

6. Would you feel comfortable sharing your beliefs about cheating with your younger sister or brother, cousin, nephew, or niece?   Yes or No?

Side Two

Chapter 6

Demystifying Math

We are going to play a game. I know the rules, but you do not. What are your chances of winning? Slim!

You are taking a math course and you have decided to only take notes and ignore the homework. You write everything down that the teacher writes on the board. These notes include information on mathematical skills, concepts, and problem-solving strategies along with examples of problems' solutions. Each night you study your notes and summarize in your own words what you learned that day. Do you have a chance to pass this course?

You certainly do! You know the "rules." That is, you know the skills, concepts, and strategies required for solving problems involving these concepts. Therefore, it is possible that you could apply these "rules" to other problems. If you do not know the "rules", you cannot apply them.

One of the most important activities students do is retain a record of all the "rules" in a journal. A journal is a separate notebook for notes only — just notes from class and students' own reflections — no homework. The teacher needs to explain to students that the journal is their own personal record of problem solving information and that its purpose is to help them learn strategies to solve real problems. Yes, the process of solving problems does entail many procedural "rules", but these rules can be looked up if they are written in their notes. The ultimate reason to study mathematics is not to memorize "rules", but to learn when and how to apply them.

Each day's "board" notes can be headlined with a title phrase or sentence. Since many students are visual learners, all the instructor's notes need to be written on the board or overhead. This lets students know that everything the teacher writes, including sample problems' solutions, goes in their journals.

The "board" notes are actually created by the teacher and class together. Students will master the information in their notes only if they rewrite them. They need to be encouraged to rewrite the notes using their own words because this transforms the information into knowledge. The process of rewriting itself is practice, and practice is necessary for mastering mathematics.

Professional athletes are enjoyable to watch because they practice. They did not master their sport the first time they played it. Professional music sounds beautiful because professional musicians practice. Musicians may experience difficulty learning how to play a song, but with practice they become adept in playing the entire song. Students may have trouble learning a concept, but with practice they too can master the whole concept. Students must not be expected to master a new idea the first time it is introduced. It takes time and practice to learn math.

Much of learning math is visual. Students often ask why something is valid, when all they need to do is look at the problem, diagram, or figure and see the reason with their eyes. For example, the expressions $(-2)^0$ and -2^0 look different and consequently they represent different numbers. Students need to look closely at the expressions to perceive that "0" in $(-2)^0$ is written above -2 whereas "0" in -2^0 is written above 2.

In order to comprehend math, students need to translate math symbols into their native language. Some concepts are best learned by actually writing math problems in words. Most people think in words, not math symbols. Part two of this book has many examples that use this process.

Students often do not know how to translate from words to math symbols. They try to interpret the entire sentence instead of translating each word or phrase separately. Students can create a word chart similar to the one below as a reference to help them learn to communicate mathematically. Of course this chart would be entered in students' journals. An entire page could be allocated for the chart so that students can add words as they encounter them.

Learning math requires no special talent. It simply requires that students be organized, take notes, and practice. Everyone can learn math!

Math Word Chart

Equal	Add	Subtract	Multiply	Divide
is	sum*	difference*	product*	quotient*
exceeds**	by**	less	exponent	cut
same	greater than	less than*	factor	separate into groups
balanced	increased by	decreased by	fraction of	fraction
			decimal of	partition
			percent of	per

*answer

**example: 8 exceeds 3 by 5
              8   =   3 + 5

***example: 2 less than 7   The meaning of "less than" requires the
              7   -   2   order of the numbers be switched.

Chapter 7

Why Learn Math?

Students are motivated to learn if they believe it is worthwhile to them. Worthwhile means that it feels good to them because it satisfies one of their basic needs as identified by William Glasser: power, fun, freedom, love, or survival. Teachers, therefore, must believe that what they are teaching is worthwhile if they expect their students to learn. Teachers are salespersons. How the information is presented is just as important as what it is. Students often ask "why learn math?" because they do not see how math fulfills any of their basic needs. One of the teacher's roles is to explain the benefits of knowing math now and in the future.

First, students will think math is worthwhile if they experience it as a language which they can use to make sense of the world. This is accomplished by teachers relating math to life and students solving real problems.

Second, mathematical ability empowers students with analytical reasoning skills and problem-solving strategies. It gives them confidence in their capacity to solve problems. Problem-solving requires that students struggle to develop alternative strategies. This is important because in real life students deal with problems which require persistence and the ability to create alternate solutions.

Third, math problem-solving skills can simplify their lives and save them time and money. For example, where is the optimum place to park a car at a mall when shopping at more than one store so that they walk a minimum distance? Let's assume that the mall is located on one street or in a linear building. Teachers can encourage students to draw a diagram to illustrate the situation. They can draw a segment whose endpoints represent the stores located furthest apart. The students trace the path walked from two different starting points to both stores and back to the starting points. Students are amazed to discover that they can park anywhere between the stores that are furthest apart and walk the same distance. Of course, they would still try to park as close to the mall as possible.

Fourth, mathematical ability gives students more earning power. The more math courses students take, the more money they will earn. Many professions, including an increasing number of technical fields, are requiring more, as well as higher level, math courses as part of their training. When students cease taking math courses, they are excluding themselves from many occupations. Students need to check the math requirements for their chosen career before they stop taking math. The longer the time span between math courses, the more difficult it is to continue taking math. If you don't use it, you lose it, definitely applies to math.

Fifth, math is used as a screening device for some jobs even though the job may not require the use of math directly. Employers such as state governments and utility companies actually give math tests to narrow the number of applicants.

Sixth, math helps students learn to deal creatively with repetition. Everyday life involves doing tasks like cooking, washing dishes, mowing, and washing clothes on regular basis. All jobs have repetitive tasks. For example, dentists, mechanics, surgeons, carpenters, and teachers all do specific tasks over and over. Dentists drill cavities, mechanics repair engines, heart surgeons operate on hearts, carpenters pound nails, and teachers stage information every day they work. Imagine that you are scheduled for heart surgery, but your surgeon decides that she is bored with operating on hearts that day, so she decides to operate on your liver instead! To put it another way, would you prefer a heart surgeon who has performed the operation many times or one with no experience?

Everyone is confronted with how to make repetitive tasks enjoyable. When students say that they are bored with math, they might mean that they are not being challenged. If this is the case, the teacher can encourage these students to pursue independent projects of their choice. However, if students are just complaining that doing math is boring, the instructor can inform them that no one is responsible for entertaining them except themselves. Life is what they make it, not what someone else does for them.

In addition to the above mentioned practical reasons, some students actually learn math for the pure enjoyment of its beauty and mental challenge. Some see beauty in all the logical reasoning involved in a solution to a problem. Others enjoy the mental challenge because they feel good after solving a problem. Most students, however, need to be frequently reminded of practical reasons for learning math.

Chapter 8

What does math have to do with life?

When students ask what a specific concept has to do with life, they may be indicating that they do not understand the concept. What they want is further explanation. When students ask what math in general has to do with life, they are really asking how math relates to life outside the classroom.

The very nature of life is mathematical. That is, nature is composed of patterns and mathematics is the language which describes them. These patterns are discovered, not created, by persons who use mathematics to communicate what they observe.

Scientists, by way of illustration, believe that there are only ninety-two naturally occurring elements which combine to form everything in the universe. Equations can be written to describe the relationship between the elements in every living and non-living thing. Not only that, but all combinations of the ninety-two elements can also be represented geometrically. Likewise, the interactions between the objects that these elements form can be modeled using mathematics.

Mathematical models are a way of making sense of the world even though they do not always describe an event exactly. Take, for example, the natural occurrence of the earth revolving around the sun. Most people believe this revolution is a simple pattern of precisely 365 1/4 days each year. Contrary to common belief, the pattern is not that simple. In fact, one earth year is exactly 365 days, 5 hours, 48 minutes, and 46.2 seconds. Since the difference between our calendar and how the pattern naturally occurs is about 11 minutes per year, the calendar needs to be adjusted from time to time. This is accomplished by dropping leap year from every year ending in "00" which is not divisible by 400. Yet even this adjustment does not describe the earth's revolution pattern precisely. Despite this fact, the calendar is a model which describes the earth's revolution very closely to its natural occurrence.

What does math have to do with life? Math is the language used to describe life events as accurately as possible. Once students have their eyes opened to this definition, they become skilled at answering the question for themselves.

Chapter 9

Gifted and Talented Students

Identifying gifted and talented students can be very difficult. As the following case illustrates, many sources need to be used to identify gifted students since special talents are often not revealed by standardized tests or grades.

A colleague once had a high school freshman algebra student tell her that he knew everything in the book one month into the school year. She checked his standardized test scores. They were average. He had obtained a B in eighth grade math. Meanwhile, he earned perfect scores on his algebra tests.

He persistently told her that he knew everything in the course. Since I had worked several years teaching and identifying gifted and talented students, she asked me to speak with him. I asked him how he learned algebra. He told me that he spent the summer teaching himself not only algebra, but also geometry, trigonometry, and calculus. He said that he had the ability to teach himself math, and that he was ready to study college calculus.

I asked his teacher to give him the algebra final exam. She agreed and he received a perfect score. That was enough evidence for the math department, administration, and his parents to allow him to independently study geometry, advanced algebra, and pre-calculus. The teachers who taught each course administered and graded his exams. He received perfect or near perfect scores on all the tests. By the end of his freshman year, he had completed first year alegebra, geometry, advanced algebra, and pre-calculus.

The math department offered AP Calculus, but this student had already taught himself this course. Fortunately, his guidance counselor was willing to pursue enrolling him at the local university as a special student for his sophomore year. It wasn't easy. This university typically did not admit high school sophomores as special students. Eventually, the school agreed to allow him to take the calculus placement exam. He received a near perfect score and was placed into second semester calculus.

During his sophomore and junior years, he continued to take math courses at the university. By the time he completed high school (a year early), he had completed almost enough courses for a math degree. If this student had had to take a typical high school math curriculum which ends, not begins, with calculus, he would have been bored and frustrated, and perhaps even dropped out of school.

This case exemplifies how difficult it is and how much work it can be to identify a person's special talent, and to make the necessary arrangements to develop that talent. While this student is exceptionally talented, all gifted and talented students need to be treated differently if they are to keep learning. Since gifted students can be enrolled in any course, all courses could have the following homework policy. Students who earn an A on two consecutive tests do not lose points for incomplete homework assignments. If a score on a test, however, brings their grade average below an A, they must complete all homework assignments towards the next test or lose points.

The intent of this policy is to encourage students to do what is necessary to learn and then move on. Nobody enjoys doing busy work. It is boring and it stifles the brain. Some students can master the material without doing every assigned problem. Many grading methods penalize gifted students by lowering their grades for incomplete homework.

Gifted and talented students need to be given opportunities to discuss with each other and the teacher how to deal with their special gifts. If possible, it is helpful for these students to get to know talented students in other areas. Gifted students often feel isolated since some students deride and others idolize them. While being idolized may initially sound good, people who are idolized are not treated as persons Everyone needs to be accepted for who they are, not what they accomplish.

Talented students need to learn how to communicate with people in a manner that does not make others feel less intelligent. People do not appreciate someone who always knows everything. Gifted students must realize that, even though they possess better reasoning abilities than most people, it is not necessary for them to exercise their analytical skills to their fullest in every conversation. When talking with people, it is best to speak to them using their "language."

Others will learn to accept gifted and talented students as persons if they do not act and talk like their special gifts make them better or more important people. Gifted students must realize that they did not create their gifts; the talents were given to them. They are responsible only for developing their given talents.

Chapter 10

Strategies for Troubled Students

Teachers are managers of students, not controllers of them. Students are responsible for controlling themselves, just as teachers control their own behavior. All too often teachers are held responsible for students' behavior. While no one is responsible for another's behavior, there are strategies that teachers can employ to help prevent discipline problems.

Following is a list of several classroom management strategies. No single strategy works forever. Therefore, teachers need to develop a repertoire of techniques. When one technique ceases to be effective, instructors can try another. Many of these strategies take extra time and effort, but discipline problems also take much time and energy.

1. Teach math courses that you feel comfortable teaching.

All students easily identify a teacher who is unable to clearly present the material. Troubled students are especially difficult to manage when they know that their teacher is not comfortable with the course content. Classroom management is much easier when a teacher teaches courses s/he is confident teaching.

2. Discuss with students how others try to control their behavior.

Students quickly grasp that some people try to control others' behavior by putting them down. Name callers, for example, try to provoke a verbal or physical reaction from their victims. In addition to attempting to control others' behavior, people put others down to elevate themselves. People who make fun of others are most likely made fun of by others. How can this circle of negativity be broken?

Persons can elevate both themselves and others by directing positive words and behaviors toward others. Such exercises can help create a circle of positivity. People who are complimented are more likely to compliment others than people who have been ridiculed.

3. Treat troubled students with dignity.

It is common today for school counselors to deal with incest, suicide, alcohol and other drug abuse, sexual assault, and depression. Kids who are dealing with these serious problems are being the best students they can be, considering their circumstances. Students who cause trouble in the classroom are themselves troubled. Teachers are not to take troubled students' inappropriate behavior personally. These students are redirecting aggression from others who have mentally or physically abused them, toward teachers.

Even though students may not initially treat their teachers respectfully, teachers can earn their respect by treating them with dignity. Troubled students may not be the best pupils, but all students have the right to be esteemed as worthy human beings. These kids especially respond positively to teachers who show an interest in them as persons. Greeting each student (with a high five, "good to see you", a handshake, etc.) as they enter or leave the classroom once a week is a simple but effective way for teachers to acknowledge each student as a person.

4. Act, don't react or overact. Have a sense of humor!

Students often misbehave to see if they can control the teacher's behavior, i.e., upset the teacher. While teachers cannot control students' behavior, instructors do control whether students control them. When students misbehave, teachers can do the unexpected by not acting upset. For example, when students pull a prank, rather than get angry, the teacher can acknowledge that they got one over her/him.

5. Get to know students outside the classroom.

Students appreciate when teachers observe them in extracurricular activities. A teacher's attendance at their activities shows them that the teacher cares about them as persons. Initially, they may reject the concern, but ultimately everyone appreciates attention.

6. Give pep talks: group and individual.

Pep talks can help manage the classroom environment. Troubled students might need a pep talk each class meeting. A pep talk can be a personal sharing of what the teacher has done to prepare for class or simply a story. Students are seldom aware of the effort teachers put into making class time enjoyable; story telling is a great way of connecting with them.

In addition to group pep talks, these students need individual positive feedback. This is physically demanding, but it is humanly possible and very rewarding for both parties. Teachers can choose one student per day to encourage, but not praise. Praise is external motivation based on achievement and is seldom internalized. Encouragement promotes self-motivation in that it makes students aware of who they are as persons by informing them of their character traits. Character traits are real-life skills that everyone can develop. These include the ability to work with others, a positive attitude, initiative, honesty, and communication skills, to name but a few. These real-life skills need to be acknowledged as important, and encouraged in order to be developed.

Besides, will students who continually receive a grade of "C" on tests accept encouragement based upon their grade? Maybe a "C" is the best that these students can achieve on tests. Perhaps these same students exhibit an enthusiasm for life that influences others to form a positive attitude. These students can be encouraged to be a positive force by making them aware of their character attributes.

Students with poor self-images often think that their self-worth is dependent upon what they achieve, not who they are as persons. Students will feel better about themselves if they are made aware of their character traits and their importance. This is true for all students, including the gifted and talented, since gifted students often believe that their self-worth is directly related to their special mental talent, not who they are as persons.

One possibility is for teachers to put encouraging character remarks in writing and give them to students. All students will show their parent(s) a note with positive remarks from their teacher. They may never have received a proficiency report before. Another possibility is to call the student at home and offer encouragement to her/him. After speaking with the student, ask to talk to a parent and give her/him the same positive feedback.

7. Acknowledge often-absent students when they attend.

Students who are often absent hate going to school for numerous reasons. One reason they dislike coming to school is because students and teachers "get on their case" about not attending school when they do attend. A colleague had a student, who was frequently absent from school, tell him that she enjoyed coming to his class because he always acted happy to see her.

8. Assign one student per class you can rely on to go get help.

Teachers can prearrange with at least one student from each class the process for getting help if a crisis arises. If an incident occurs, teachers need to remain calm and not become irate or aggressive. Students not involved in the incident are to leave the classroom. The teacher needs to tell an administrator about a crisis as soon as possible.

9. Develop a disruptive-student policy.

Teachers need to have their own set of measures for students who are so disruptive that they need to be asked to leave class. Following is one example of a teacher's set of procedures for dealing with disruptive students.

First Offense: Hold a meeting with the student. The teacher needs to tell the student that they are meeting to solve the problem, not to punish her/him.

Second Offense: Call the parent (the one that the student wrote on the index card the first day of class) to discover how much the parent is a part of the problem and how much the parent is willing to be part of the solution. If possible, have the student call a parent in your presence to explain the situation. The teacher then talks to the parent, after the student .

Third Offense: Hold a meeting with the student and counselor and/or dean of students.

Fourth Offense: Hold a conference with the parent(s), student, counselor, and/or dean of students.

At each step, teachers need to define their purpose as problem solving and not as punishment. Teachers must be specific, stick to the facts, and be persistent with the facts concerning this student. Do not argue with the student or parent. Never discuss other students. If the conversation is sidetracked, the teacher redirects to its purpose. If teachers discover that they made a mistake, they need to apologize. This lightens the mood and reminds everyone that teachers are human.

Classroom management strategies help create a learning environment; they are not techniques for controlling students' behaviors. Students are responsible for their own actions. While these techniques might alleviate behavior problems, they will not necessarily affect students' academic performance. Each individual student chooses to learn or not to learn.

Part Two

Reality

Based

Instruction

Chapter 11

What is Math?

Math is a universal foreign language which is used to describe patterns. It is universal in that all countries use the same symbols with the same meanings. It is a foreign language because the symbols need to be translated into the students' native language in order for them to understand the meaning of the symbols. The symbols are used to describe patterns. Students are amazed to discover that people worldwide use the same symbols, including the variables "x" and "y". In summary, the following could be written on the board and into students' journals.

Math is a <u>universal</u> <u>foreign language</u> which is used to <u>describe patterns</u>.

It is <u>universal</u> in that all countries use the same symbols with the same meanings.

It is a <u>foreign language</u> because the symbols need to be translated into the students' native language in order for them to understand the meaning of the symbols.

The symbols are used to <u>describe patterns</u>.

If students have difficulty understanding math symbols, teachers can ask them to translate the math symbols into words. Upon translation, lack of understanding dissipates, and students discover that using symbols is a short cut. Once students understand that math symbols are merely short cuts for representing words, they are ready to discover how math is used to describe patterns. Following is an activity whereby students generate a formula to describe the pattern that exists between their grade levels and their ages.

Teachers begin this activity by drawing a chart on the board with age as a heading for one row and grade as the heading for the other. Next, students are asked how old they were when they entered grade one. Some say five and others say six years old. The students soon discover that a formula cannot be generated unless an assumption is made about the students' age when entering grade one. This shows students first hand the need for assumptions in math.

After some discussion, the class agrees to assume that students are 6 years old when entering grade 1, since six is the prevailing age for first graders. The chart can then be completed as shown below.

Assumption: Students are six years old in first grade.

age	6	7	8	9	10	......	a
grade	1	2	3	4	5	......	g

Students are quick to see the pattern. One student might say that a person's grade level equals their age minus five, and another student might say that a person's grade level plus five equals their age. Since they know that symbols can be used to represent words, they will want to substitute "g" for grade level and "a" for age. In summary, the following could be written on the board.

grade equals age minus five   grade plus five equals age

$$g = a - 5 \qquad\qquad g + 5 = a$$

These formulae make a lot more sense to students than $y = x - 5$ or $y + 5 = x$. Math makes sense when students know what the letters represent. The challenge for math teachers is to give meaning to variables whenever they are used. The chapter entitled, "Operations on Variables," illustrates how this can be done.

In addition to defining mathematics, teachers can describe the different branches of math. Algebra can be defined as a language which is used to solve patterns (equations) for an unknown (represented by a letter which is called a variable). Geometry involves the study of the measure of earth objects. Trigonometry is the study of the measure of similar triangles. If the triangles are not similar, they cannot be compared because their sides are not proportional. Calculus is the study of methods for calculating any quantity (e.g., area and volume of objects of all shapes) and rates of change (e.g., velocity and acceleration). Probability is the study of the likelihood that events will occur. Statistics is the study of the collection, analysis, classification, and interpretation of information. In summary, the following could be written on the board and into students' journals.

Algebra is a language used to solve equations for an unknown (variable).

Geometry is a language used to measure earth objects.
Geo means earth and metry means measure.

Trigonometry is a language used to measure <u>similar</u> triangles.
Tri means three, gon means angle, and metry means measure.

Calculus is a language used to calculate fixed and changing quantities (rates of change).

Probability is a language used to predict the likelihood that events will occur.

Statistics is a language used to collect, analyze, classify, and interpret information.

Math makes more sense to students if they know what it is they are studying. Most students do not understand what math is because most math books do not define it. Students will be much more comfortable learning if teachers define mathematics and its different branches in all courses.

Chapter 12

A Process for Solving Problems

In real life, acknowledgment of a problem is essential and often the most important step to its solution. Math teachers are very fortunate since math problems are readily acknowledged by students! Once a problem is acknowledged, strategies can be developed to solve it. Teachers need to provide a variety of experiences where students can create a repertoire of problem-solving strategies.

I introduce problem-solving to my students by illustrating how my wife and I resolved a real problem using the same problem-solving process that they will utilize to solve math problems. A real problem is one whose solution is not known prior to the solving. Here was our problem.

> My wife needed further schooling for her career. Neither of us had a job in or nearby the city where this school is located. It was very difficult to secure jobs in or around this city. Where could I work and where could we live together so that my wife could attend this school?

Step 1: Identify the Problem.

A problem must be identified before it can be solved, but identifying a problem can be very difficult. It took us quite a while before we realized that we had a distance problem. Since we had only one vehicle, we were fixated on the idea that we had to live in this city. In order to explore all possible solutions, we proceeded to solve this problem, assuming that another car might need to be purchased. Once we identified that the problem was distance, we applied the remaining steps of the problem-solving process to generate possible solutions.

Even though algebra students solve many distance problems, they, too, have difficulty identifying this real-life problem as distance.

Step 2: Develop Specific Strategies. Simplify problem, if possible.

We assumed that neither of us wanted to commute more than thirty miles one way. (I mention to my students that math could not exist without assumptions, i.e., axioms or postulates.) There were two distance problems to

38

solve. Where could I work, and where could we live together that would satisfy the above assumption?

Draw a picture, construct a table, or plot a graph.
Calculate by hand, calculator, or computer.

The class determines that to solve this problem we must have obtained a state map. I ask the students if they have any ideas about what we drew on the map to help answer our two questions. They say we drew a circle with a radius of thirty miles around the city. I ask them which question the circle helps to answer. Students state that the area within the circle represents the many possible locations where we could live. Next, students are asked what we drew to represent where I could work. After some deliberation, students say we drew a concentric circle with a radius of sixty miles.

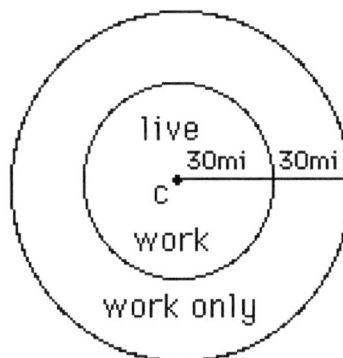

c = city where school is located

Step 3: Translate Problem into Math. Solve it.

Which did we acquire first, a place to live or a job for me? Since it is more difficult to secure a job than a place to live, students respond that I found a job first. Once I obtained a job within the sixty mile radius of the school, we found a place to live in one of the locations that would allow both of us a commute of less than or equal to thirty miles. Some students are amazed that two people could spend their days sixty miles apart and still live together.

In the process of exploring possible solutions using this mathematical model, we discovered a solution that required us to alter our initial assumption that neither of us commute more than thirty miles to work and school. Both of us were offered employment at the same place in a city located seventy miles from the school. We were not looking for a job for my wife, but this meant that we would need just one car because she could attend school part-time in the evening.

Step 4: Check Solution(s) Mathematically.

Are the circles drawn properly on the map? Is the distance from our house to our jobs less than or equal to thirty miles?

Step 5: Answer Question(s). Choose a Solution.

Once we accepted the jobs, we chose a place to live. Since my wife would only need to drive to school one evening per week, and we worked five days per week, we chose to live 25 miles from our workplace. The total distance traveled to and from work each week was 250 miles (50 miles/day times 5 days). My wife drove 45 miles to school once per week for an additional distance of 90 miles (45 miles one way times 2). The total commuting distance traveled per week was 340 miles (250+90). While the initial assumption was slightly altered, we still drove daily less than thirty miles to work. Our total distance traveled per week was significantly less than the maximum of 600 miles possible under the original assumption (30 miles/day one way times 2)(5 days)(2 cars).

Step 6: Check Solution for Practicalness.

Even though the place where we chose to live was less than thirty miles from our jobs, we checked how long it would take to drive to our workplace and to the school. A given distance can take different times to drive, dependent upon the amount of traffic and the number of required stops.

Students are now aware that the problem-solving process, which they will use to solve math problems, can be applied to life outside the classroom. In order to give students an opportunity to apply the process, the following problem can be solved in groups of two.

> A person is in charge of hiring officials for the men's and women's NCAA basketball tournaments. There are sixty-four teams in their single elimination tournaments and three officials referee each game. How many officials need to be hired for each tournament?

Step 1: Identify the Problem.

Students read the problem and identify strategies that will help solve it. They do not need to be concerned with the question at this time. Students discover that this problem involves counting the number of games.

Step 2: Develop Specific Strategies.
   a. Simplify problem if possible.
   b. Draw a picture, construct a table, or plot a graph.
   c. Calculate by hand, calculator, or computer.

Students read the problem again to see if it can be simplified. (Once again, they do not need to be concerned with the question at this time.)

Groups can simplify the problem by changing the total number of teams to a smaller number in order to ascertain the pattern between the number of teams and the number of games. The teacher can show them how to record the information in a table similar to the one below. Those who simplify the problem usually discover the pattern before completing the table for sixty-four teams.

#teams	2	4	8	16	32	64
#games	1	3	7	15	31	63

Other groups will make a table listing the number of games played at each round of the tournament, and then add up the number of games (32 + 16 + 8 + 4 + 2 + 1 = 63) as illustrated in the following chart.

round	first	second	third	fourth	fifth	sixth
#teams	64	32	16	8	4	2
#games	32	16	8	4	2	1

A few groups may deduce that the number of games is one less than the number of teams since one team wins. In each of the tournaments, sixty-three games will be played because if one team is the winner, sixty-three teams must lose.

Step 3: Translate Problem into Math. Solve it.

$$g = \text{number of games}$$
$$t = \text{number of teams}$$

$$g = t - 1$$
$$g = 64 - 1$$
$$g = 63$$

Step 4: Check Solution(s) in the Equation(s).

In this case, the solution was calculated by substituting it into the equation.

Step 5: Answer Question(s). Choose a Solution.

Now the question is answered! In many cases the question can be answered only after an equation is solved. And the solution to the equation is not necessarily the answer to the question as in this case. The question concerns how many officials need to be hired for sixty-three games. Given the fact that three officials are required for each game, a total of one hundred eighty-nine ($63 \times 3 = 189$) officials need to be hired.

Step 6: Check Solution for Practicalness.

Does the solution make sense? Do 189 *different* officials need to be hired? Students can discuss the fewest number of different officials needed. Of course it depends on the number of locations and the times at which the games are played. If all games are played at one site at different times, only three different officials are needed. However, they would be three tired officials!

After each group has answered the question, they could share their different strategies with the class. The teacher might want to highlight how the problem could be solved by simplifying. Students will see how simplifying, when possible, helps clarify the problem and reduces the amount of work.

As a special project, a student could write the NCAA to get a diagram of both tournaments that gives the number of sites for each round. Students could use this information to propose how many different officials realistically need to be hired.

Using real problems to introduce a process for solving math problems lets students know that they can apply the process to their own difficulties. It is recommended, however, that they make one additional check. It is important to check personal decisions by filtering them through the heart. This way decisions will be verified both rationally and emotionally before being acted upon.

Students will feel confident in their capacity to solve problems if they experience a variety of problem-solving strategies. Students can write these strategies in a separate section of their journals, as they discover them, so that they can refer to them as needed. Below is a sample list of some problem-solving strategies.

Problem Solving Strategies

1. Read the entire passage as many times as needed in order to identify the problem.

2. Do not worry about answering the question until after solving the equation(s).

3. Translate the problem from English into math symbols, a word or phrase at a time.

4. Look up words in your journal, book, or dictionary.

5. Simplify the problem:
   First solve the problem by substituting a number for a variable. Or first solve the problem by substituting a smaller rounded number for a larger number.

6. Draw a picture, table, chart, or diagram.

7. Decide whether to calculate by hand, calculator, or computer.

8. Plot a graph by hand or on a graphics calculator.

9. Remember to answer the question.

10. Check solution(s) for practicalness by estimating.

11. Be persistent. Try another strategy.

Chapter 13

Fractions are Fair — Decimals are Decent!

Advanced, as well as basic, math students experience difficulty working with fractions. Fractions are difficult to understand because they can be described in many ways. Consequently, teachers of all math courses can assume that a summary of fractions is welcomed by most students.

This chapter is a review of both fractions and decimals. The comparison of fractions to decimals helps students understand how both are used to represent part(s) of one whole. The following material is designed to be discussed with students over several class periods after they are comfortable working with both fractions and decimals.

Fractions are Fair!

1. A fraction is part of one whole thing.

2. A fraction is a number between -1 and 1.

3. A fraction $= \dfrac{\text{number of "equal parts" remaining}}{\text{number of "equal parts" in one whole thing (size)}}$

The denominator tells the relative size of the parts.
The more parts there are in one whole thing, the smaller the parts.

4 parts **2 parts**

smaller parts / larger parts

"Equal parts" guarantees that the whole thing is divided fairly.

not a fraction **a fraction**

not fair / fair

4. A fraction $= \dfrac{\text{numerator}}{\text{denominator}}$

The _d_enominator is located _d_own under and is the _d_ivisor. All three words begin with the letter "d". Students can easily remember this, but they still need to know how to translate a fraction into a division problem. With a little imagination, the following illustration will help everyone remember how to write a fraction as a division problem.

The division symbol $\overline{}$ can be made to look like a cap ___$\overline{}$. The brim of a cap is also called a visor. You guessed it! The divisor goes on the visor.

$$\underline{\textbf{divisor}}\overline{}$$
$$\textbf{visor}$$

5. A fraction is a quotient, which is a decimal.

$$\textbf{fraction} = \overline{\vert\textbf{quotient}}$$

$$\textbf{fraction} = \overline{\vert\textbf{decimal}}$$

$$\frac{1}{4} = 4\overline{)1.00}^{\,.25}$$

The decimal point goes at the end of the whole number 1, just as a period is placed at the end of a sentence. The words decimal and fraction are synonyms, therefore, the decimal .25 can be expressed as a fraction.

$$.25 = \frac{25 \text{ equal parts remaining}}{100 \text{ equal parts in one whole thing}}$$

Since decimals are another way of expressing fractions, all of the above explanations describe decimals. _Decimals are decent_ means that all parts are equal. However, the decimal parts in one whole thing are only multiples of ten, (10, 100, 1000, etc.).

Improper Fractions

When the numerator is greater than the denominator, the fraction is called *improper* because it does <u>not</u> *fit* the definition of a fraction.

$\dfrac{7}{4}$ is an improper fraction because it is not part of one whole.

$\dfrac{7}{4}$ equals $1\dfrac{3}{4}$ which is more than one whole object.

$$\frac{7}{4} = 4\overline{)7}\quad \underline{\textbf{1 whole and 3 parts remaining}} = 1\ \frac{3}{4}$$

The number 1 can be expressed many ways.

$$1 = \frac{1}{1} \qquad\qquad 1 = \frac{4}{4} \qquad\qquad 1 = \dfrac{\frac{3}{5}}{\frac{3}{5}}$$

Mixed Numbers

$1\dfrac{1}{4}$ is called a mixed number because it is composed of a whole and a fraction.

$$1\frac{1}{4} \quad=\quad 1 \quad+\quad \frac{1}{4}$$
$$\text{mixed number}\quad\text{whole}\quad\text{fraction}$$

$$1\frac{1}{4} \quad=\quad \frac{4}{4} \quad+\quad \frac{1}{4}$$

$$1\frac{1}{4} \quad=\quad \frac{5}{4}$$

Any number can be divided by 1, and its value remains the same.

$$0 \div \frac{1}{1} = \dfrac{0}{\frac{1}{1}} = 0 \qquad 5 \div \frac{4}{4} = \dfrac{5}{\frac{4}{4}} = 5 \qquad \frac{2}{7} \div \frac{4}{4} = \dfrac{\frac{2}{7}}{\frac{4}{4}} = \frac{2}{7}$$

Reducing Fractions.

Fractions are reduced by dividing by an expression of one other than $\dfrac{1}{1}$.

$$\frac{2}{6} \div \frac{2}{2} = \frac{1}{3} \qquad\qquad \frac{6}{21} \div \frac{3}{3} = \frac{2}{7} \qquad\qquad \frac{12}{16} \div \frac{4}{4} = \frac{3}{4}$$

Increasing the Denominator
Any number can be multiplied by 1, and its value remains the same.

$$\frac{3}{4} \times \frac{2}{2} = \frac{6}{8} = \frac{3}{4} \qquad \frac{3}{4} \times \frac{4}{4} = \frac{12}{16} = \frac{3}{4} \qquad \frac{3}{4} \times \frac{5}{5} = \frac{15}{20} = \frac{3}{4}$$

The Four Operations

The four basic operations in math get physical with numbers and variables just as medical operations involve physical actions by a surgeon. Addition *combines* like objects. Subtraction *gives or takes away* like objects. Multiplication of <u>fractions</u> and <u>decimals</u> *divides* a part(s) into smaller equal sized parts. Division of <u>fractions</u> and <u>decimals</u> *gives the number of parts* into which a part(s) is partitioned.

Subtraction of fractions gives or takes away a part(s).

People cannot give away something that they do not possess, but they can alter what they have. For example, people cannot give away a dime if they only have a quarter. They can give away a dime if they exchange the quarter for a combination of dimes, nickels, and/or pennies. The above situation can be depicted mathematically in the following manner.

A quarter is $\frac{1}{4}$ of a dollar or twenty-five cents, $\frac{25}{100} = \frac{1}{4}\left(\frac{25}{25}\right)$.

A dime is $\frac{1}{10}$ of a dollar or ten cents, $\frac{10}{100} = \frac{1}{10}\left(\frac{10}{10}\right)$.

$$\frac{1}{4}\left(\frac{25}{25}\right) = \frac{25}{100}$$
$$- \frac{1}{10}\left(\frac{10}{10}\right) = - \frac{10}{100}$$
$$\frac{15}{100}$$

Converting a quarter and a dime to pennies (like objects) is called getting a common denominator. The common denominator makes the parts equal (alike) so that operations can be performed. The answer $\frac{15}{100}$ can be reduced by dividing by $\frac{5}{5}$, thus $\frac{15}{100} \div \frac{5}{5} = \frac{3}{20}$. In other words, $\frac{15}{100}$ of a dollar is the same as $\frac{3}{20}$ of a dollar.

Addition of fractions combines equal size parts.

A total quantity can be calculated only if the objects are alike. Everyone knows that a quarter and a dime make a total of thirty-five cents. What everyone does not realize is that adding a quarter and a dime requires that the coins be made alike by converting them to pennies.

A quarter is $\frac{1}{4}$ of a dollar or twenty-five cents, $\frac{25}{100} = \frac{1}{4}\left(\frac{25}{25}\right)$.

A dime is $\frac{1}{10}$ of a dollar or ten cents, $\frac{10}{100} = \frac{1}{10}\left(\frac{10}{10}\right)$.

$$
\begin{array}{rcl}
\frac{1}{4}\left(\frac{25}{25}\right) & = & \frac{25}{100} \\[2mm]
\frac{1}{10}\left(\frac{10}{10}\right) & = + & \frac{10}{100} \\[2mm]
& & \frac{35}{100}
\end{array}
$$

The answer $\frac{35}{100}$ can be reduced by dividing by $\frac{5}{5}$, thus $\frac{35}{100} \div \frac{5}{5} = \frac{7}{20}$.

In other words, $\frac{35}{100}$ of a dollar is the same as $\frac{7}{20}$ of a dollar.

Multiplication of fractions divides a part(s) into smaller equal parts.

Half of a dime ($\frac{1}{10}$ of a dollar) is a nickel ($\frac{1}{20}$ of a dollar). The dime is partitioned into two smaller equal parts; each part is one-twentieth of a dollar.

Half of a dime is a nickel.

$$\frac{1}{2} \ \times \ \frac{1}{10} \ = \ \frac{1}{20}$$

Division of fractions *generates the number of parts* into which a given part(s) is separated.

A quarter divided by a nickel generates the number of nickels (twentieths) in one quarter (one fourth). Everyone knows that there are five nickels in a quarter. This means that there are five twentieths ($\frac{5}{20}$) in one fourth ($\frac{1}{4}$) and that five twentieths reduces to one-fourth, $\frac{5}{20} \div \frac{5}{5} = \frac{1}{4}$. The following illustrates how five is calculated using math notation. A quarter is $\frac{1}{4}$ of a dollar and a nickel is $\frac{1}{20}$ of a dollar.

A quarter divided by a nickel is the number of nickels in a quarter.

$$\frac{1}{4} \quad \div \quad \frac{1}{20} \quad = \quad 5$$

A division problem can be rewritten as a fraction. The <u>d</u>ivisor is what comes after the words "<u>d</u>ivided by" and is located in the <u>d</u>enominator.

$$\frac{1}{4} \div \frac{1}{20} = \frac{\frac{1}{4}}{\frac{1}{20}}$$

This fraction can be multiplied by 1 expressed as $(\frac{20}{20})$ and its value will not change. The fraction $(\frac{20}{20})$ is used to represent 1 because 20 is the common denominator of 4 and 20.

$$\frac{\frac{1}{4}}{\frac{1}{20}} \times \left(\frac{20}{20}\right)$$

The number 20 can be divided by $1(\frac{20}{1})$ and its value will not change.

$$\frac{\frac{1}{4} \times \frac{20}{1}}{\frac{1}{20} \times \frac{20}{1}} = \frac{\frac{20}{4}}{\frac{20}{20}} = \frac{5}{1} = 5$$

Decimals are Decent!

1. A decimal is part of one whole thing.
2. A decimal is a number between -1 and 1.
3. A decimal is located to the right and integers to the left of the decimal point.

integer . decimal

4. A decimal is a fraction whose denominator is a multiple of 10.

$$.\frac{1}{10} \qquad \frac{1}{100} \qquad \frac{1}{1000}$$

. tenths   hundredths   thousandths

. dimes   pennies   no money equivalent
          or cents

Subtraction of decimals is giving or taking away a multiple of tenths:
$\frac{1}{10}, \frac{1}{100}, \frac{1}{1000},$ **and so forth.**

Let's assume a person has fifty cents and owes someone twenty-five cents. In the decimal system there are only dimes and pennies since the value of each place is worth a multiple of tenths ($\frac{1}{10}$). Fifty cents is written as .50 (5 dimes and 0 pennies). Twenty-five cents is expressed as .25 (2 dimes and 5 pennies). In order to give away five cents, one of the dimes must be *exchanged* for ten pennies. Many students learn this procedure as borrowing. Borrow means to receive with the intent to return. In math, there is no borrowing. The dime is *exchanged* for ten pennies with no intent of returning the dime.

$$
\begin{array}{ll}
.50 = & .4^{1}0 \\
\underline{-.25} & \underline{- .25} \\
& .25
\end{array}
$$

Addition of Decimals is combining multiples of tenths.

Let's calculate mathematically the total amount of five dimes and three quarters. Since it makes sense to add only like objects, the dimes and quarters need to be expressed as pennies (hundredths). Five dimes is fifty cents or .50 (5 dimes and 0 pennies). Three quarters are worth seventy-five cents or .75 (7 dimes and 5 pennies). There is a total of 12 dimes (5+7), ten of which can be *exchanged* for one whole dollar.

$$
\begin{array}{l}
.50 \\
\underline{+.75} \\
1.25
\end{array}
$$

The decimal points are aligned when adding and subtracting so that like things are combined or given away; tenths are combined with other tenths or given away, hundredths are combined with hundredths or given away, and so forth. Aligning the decimal point is the same as finding a common denominator with fractions. Let's rework the above problem using both fractions and decimals to see how aligning the decimal point gets a common denominator.

Let's use .5 to represent five dimes and .75 to represent three quarters.

$$.5 \quad = \quad .50 \qquad\qquad \frac{5}{10}\left(\frac{10}{10}\right) = \frac{50}{100}$$

$$+.\underline{75} = +.\underline{75} \qquad\qquad +\frac{75}{100} = +\frac{75}{100}$$

$$1.25 \qquad\qquad\qquad \frac{125}{100} = 1.25$$

Comparing addition of decimals to fractions shows students that adding a zero to the end of a decimal is actually multiplying by 1 expressed as $\left(\frac{10}{10}\right)$.

$$.5 = .50 \text{ because } \frac{5}{10}\left(\frac{10}{10}\right) = \frac{50}{100}$$

Multiplication of decimals divides a part(s) into smaller equal parts.

Let's compare multiplication of decimals with multiplication of fractions. As discussed with multiplication of fractions, half of a dime is a nickel. The dime (one-tenth of a dollar) is partitioned into two smaller equal parts; each part being one-twentieth of a dollar. One half $\left(\frac{1}{2}\right)$ equals the quotient .5. A dime $\left(\frac{1}{10}\right)$ equals the quotient .1. A nickel $\left(\frac{1}{20}\right)$ equals the quotient .05.

Half of a dime is a nickel.
$$\frac{1}{2} \quad \times \quad \frac{1}{10} \quad = \quad \frac{1}{20}$$

$$.5 \quad \times \quad .1 \quad = \quad .05$$

$$\frac{5}{10} \quad \times \quad \frac{1}{10} \quad = \quad \frac{5}{100}$$

Comparing multiplication of decimals to multiplication of fractions shows students that the decimal places are added when multiplying because the denominators are also multiplied.

Division of decimals *generates the number of parts* **into which a given part(s) is separated.**

Let's compare division of decimals with division of fractions. As discussed with division of fractions, a quarter divided by a nickel generates the number of nickels (twentieths) in one quarter (one fourth). Everyone knows that there are five nickels in a quarter. A quarter ($\frac{1}{4}$) equals the quotient .25. A nickel ($\frac{1}{20}$) equals the quotient .05.

A quarter divided by a nickel is the number of nickels in a quarter.

$$\frac{1}{4} \quad \div \quad \frac{1}{20} \quad = \quad 5$$

$$.25 \quad \div \quad .05 \quad = \quad 5$$

A division problem can be rewritten as a fraction. The <u>d</u>ivisor is what comes after the words "<u>d</u>ivided by" and is located in the <u>d</u>enominator.

$$.25 \div .05 = \frac{.25}{.05}$$

This fraction can be multiplied by 1 expressed as ($\frac{100}{100}$) and its value will not change. The fraction ($\frac{100}{100}$) is used to represent 1 because the common denominator of .25 ($\frac{25}{100}$) and .05 ($\frac{5}{100}$) is 100. Comparing division of decimals to division of fractions shows students why the decimal places are moved when dividing decimals. The division problem can be rewritten as a fraction and then multiplied by one expressed as a multiple of ten.

$$\frac{.25}{.05} \left(\frac{100}{100} \right) = \frac{25}{5} = 5$$

This ends the review of fractions and decimals.

Chapter 14

Division by Zero Gives Meaning to Life!

Division by zero is an important topic for every student of math. Both algebra and calculus students need to understand division by zero in order to comprehend the meaning of undefined slope and the concept of a function. Calculus students also encounter division by zero in limit problems. Yet, students of all ability levels in all math courses have difficulty discerning whether $\frac{0}{6}$ or $\frac{6}{0}$ is zero.

In the last chapter, students learned how a fraction can be converted into a division problem. The division symbol ⌐ can be made to look like a cap ‒‒‒⌐. The brim of a cap is also called a visor. The number in the <u>d</u>own position is the <u>d</u>ivisor and is placed on the visor.

$$\frac{0}{6} = 6\overline{)0}$$
$$\qquad\qquad \frac{6}{0} = 0\overline{)6}$$

The first division problem, $\frac{0}{6}$, is asking if zero things can be placed into six groups. Since zero represents nothing, nothing can only be separated into zero groups. I tell my students if they can make something from nothing to see me immediately after class!

The second division problem, $\frac{6}{0}$, is asking if six people can be partitioned into zero groups. Six people can be divided into three groups of two people. Six people can be partitioned into two groups of three people. Six people can also be separated into one group, but they cannot be divided into zero groups. Division by zero is impossible. Mathematically speaking, people have meaning in life merely by existing since they cannot be reduced to zero!

Chapter 15

How can Two Negatives be Positive?

There are many associations of negative and positive numbers in daily life. Positive is associated with earnings, gambling wins, profits, a gain of yards in football, and the directions up and right. Negative is associated with debts, gambling losses, deficits, a loss of yards in football, and the directions down and left. Any of these associations can be used to explain the rules for adding positive and negative numbers. Following is a sample presentation.

Addition Rules

Example 1. A person gets paid $200 one week and $150 the next week for total earnings of $350.

Earnings plus Earnings equals Total Earnings
    200   +   150   =   350

<u>Positive</u> plus <u>Positive</u> equals <u>Positive</u>

Example 2. A person owes $300 on master card and $450 on VISA for total credit card debts of $750. These numbers are actually negative since they represent moneys the person owes.

Debt plus Debt equals Total Expenses
-$300 + -450   =   -$750

<u>Negative</u> plus <u>Negative</u> equals <u>Negative</u>

Example 3. In the previous two examples, the person had total earnings of $350 and total credit card expenses of $750. In order to ascertain whether the person is in a positive or negative situation, the earnings need to be compared to debts. If the earnings are more than the expenses, the person is in a positive situation. The person is in a negative situation if the expenses are greater than earnings.

Earnings plus debts
    350   +   -750

Step one.   Debts (750) are more than earnings (350),
            so the answer is negative(-).

Step two.   Subtract 350 from 750 to compute the amount of debt (75-350=400).

            350 + -750 = -400

            Positive plus Negative or a Negative plus Positive
            Step one. Decide whether there are more positives or negatives.
                       Write the appropriate sign.
            Step two. Subtract the smaller from the larger number (disregard
                       the signs).

Before explaining the subtraction rules, teachers need to discuss with students another association of positive and negative. A positive sign placed in front of a number does not change the number's value. For example, if a number is written as 2, it can be rewritten as +2 without changing the number's value of positive two. Students agree that -6 and +(-6) both have a value of negative six. (Teachers need to inform students that parentheses are written only to make reading easier.) A positive sign does not alter the existing state or value of a number.

A negative sign <u>does</u> change the existing state or value of a number. A negative sign can also be translated as "the opposite of what exists." Thus, -6 can be read "the opposite of positive six" which is of course negative six. The words "the opposite of" give meaning to expressions like -(-6), which is read "the opposite of negative six" which equals positive six. Following is a sample presentation of the subtraction rules.

Subtraction Rules.
Step one.   Translate the subtraction problem into an addition problem.
            A positive sign can be inserted without changing the problem.
            A negative sign can be read "the opposite of."
Step two.   Follow the addition rules.

Example 1.   5 - 6   This is read five <u>plus</u> <u>the opposite of</u> six.
Step one.   5 + (-6)   A plus sign is inserted.
Step two.   -1   Follow the addition rules.

Example 2.   5 - (-6)   This is read five <u>plus</u> <u>the opposite of</u> <u>negative six.</u>
Step one.   5 + 6   The opposite of negative six is positive six.
Step two.   11   Follow the addition rules.

Example 3.   -5 - 6   This is read negative five <u>plus</u> <u>the opposite of</u> six.
Step one.   -5 + (-6)   A plus sign is inserted.
Step two.   -11   Follow the addition rules.

Example 4.   -5 - (-6)   This is read negative five <u>plus</u> <u>the opposite of</u>
                            <u>negative six</u>.
Step one.   -5 + 6   The opposite of negative six is positive six.
Step two.   1   Follow the addition rules.

The multiplication rules can be explained using the associations that positive does not change the state in which something exists and that negative means the opposite of what exists. The following procedure is especially effective for showing students how two negatives can be positive when multiplying.

Classroom lights can be used to illustrate the multiplication rules. During this activity, the room must be in the same original state prior to demonstrating each case. Because most classrooms have lights on when students enter, let's assume that "lights on" is the original state of the room. During this activity, a positive sign represents "lights on" or "do not flick the light switch," since neither of these conditions alters the existing state of the room. A negative sign stands for "lights off" or "flick the switch," because these actions change the existing state of the room to its opposite condition. Each of the following multiplication problems is actually worked by the teacher flicking lights on and off.

Multiplication Rules
The lights must be on prior to demonstrating each case.

+ equals "lights on"   - equals "lights off"
+ means "do not flick switch"   - means "flick switch"

Case 1.   + times +
        (lights on)(do not flick switch)   equals   "lights on"
        (Positive)(Positive)   equals   Positive

Case 2.   + times -
        (lights on)(flick switch)   equals   "lights off"
        (Positive)(Negative)   equals   Negative

Case 3.   - times +
        (flick switch)(do not flick switch)   equals   "lights off"
        (Negative)(Positive)   equals   Negative

Case 4.   - times -
        (flick switch)(flick switch)   equals   "lights on"
        (Negative)(Negative)   equals   Positive

The teacher could discuss with the class other dichotomous situations, e.g., opening and closing doors or windows, to discern if two opposites performed consecutively results in the original state (positive). While this activity does not explain *why* two negatives are positive, it does illustrate *how* two negatives can be positive when multiplying. Since division is multiplication, that is, dividing is multiplying both the numerator and denominator by the reciprocal of the denominator, the division and multiplication rules are the same. Students could look at the above cases to summarize the multiplication and division rules in the following manner.

Multiplication and Division Rules
Same Signs equal Positive
Different Signs equal Negative

All of the above examples for addition, subtraction, and multiplication, including the explanations in English, are written by students into their journals.

Chapter 16

Operations on Variables

A variable is a letter or group of letters that represents something. For example, "ft" usually stands for feet. Since a ft times 1 and a ft divided by 1 equals a ft, the symbol "ft" can be written as "1ft" or as "$\frac{ft}{1}$". Any variable times one or divided by one equals that variable. The symbol "ft" can be read "feet to first power" which is the same as "ft^1". Likewise, any variable can be expressed with an exponent of one. These facts will be useful when exploring the addition, subtraction, multiplication, and division of variables.

$$\text{variable} = 1\text{variable} \qquad \text{variable} = \frac{\text{variable}}{1} \qquad \text{variable} = \text{variable}^1$$

$$x = 1x \qquad\qquad x = \frac{x}{1} \qquad\qquad x = x^1$$

Everyone knows that a person who has only five apples cannot give away two bananas. This situation could be described using math symbols as 5a-2b where "a" represents an apple and "b" stands for a banana. If students think of variables as labels when subtracting, they will understand why the expression 5a-2b cannot be simplified.

Can 5a+2b be simplified? Some students will say that 5a+2b=7f. That is, five apples plus two bananas can be added to give seven fruit. However, the label fruit does not state what kind of fruit is being added. The seven fruit could be grapes and oranges. The expression 5f+2f adds fruit but not necessarily apples and bananas. Other students will say that 5a+2b=7ab. That is, five apples plus two bananas equals seven apples and bananas (5a+2b=7ab). However, "ab" cannot represent apples and bananas because it represents one object. The letters "ab" could stand for apple banana juice which is one liquid. The expression 5ab+2ab can represent the addition of jars of apple banana juice, not apples and bananas.

The teacher and students can explore together how groups of letters represent one object. For example, the letters "cf" and "bf" could stand for corn flakes and bran flakes respectively; the letters "mm" could represent m&m's™. This discussion helps clarify why the expression 3cf + 7bf - 4mm cannot be simplified.

The letters "mm" can be rewritten as m^2 because an exponent indicates the number of times the base is a factor. In order to help students visualize exponents, variables can be thought of as dimensions. Area is measured in square units (exponent 2) because it is composed of 2 dimensions. A space 5ft by 3 ft occupies 15 square ft which can be written as $15ft^2$ ($ft^1 \times ft^1 = ft^2$). Volume is measured in cubic units (exponent 3) because it is composed of 3 dimensions. A space 5ft by 3ft by 8ft occupies 120 cubic ft which can be written as $120ft^3$ ($ft^1 \times ft^1 \times ft^1 = ft^3$).

In the book *On the Shoulders of Giants*, Thomas F. Banchoff describes four dimensional space as it is used to model molecules. "The atoms that make up a molecule can be represented by small spheres of different radii. Each sphere requires three coordinates to specify its center (x,y,z) and one coordinate for the radius(r)." Thus, the configuration of a molecule is described in terms of the four dimensions x, y, z, and r.

A dimension is an appropriate way of visualizing a variable when multiplying and dividing. A variable can be thought of as a dimension when adding and subtracting in that only equal dimensions can be added and subtracted, but it can also be thought of as a label which represents an object.

The word <u>simplify</u> means to rephrase an expression into the fewest number of terms by performing operations in a specific order. An expression in math is a phrase of terms instead of words. Terms are factors separated by a + or - sign. So that everyone gets the same fewest number of terms, mathematicians worldwide agree to perform operations in a specified order. These are referred to as the order of operations.

Simplify:	Rephrase an expression into the fewest number of terms using the order of operations.
Expression:	A phrase of terms.
Terms:	Factors separated by a + or - sign.
Factors:	Numbers and variables that make up a product.
Coefficient:	Number before a variable.
Variables:	Labels when adding and subtracting. Dimensions when multiplying and dividing.

<u>Simplify</u> means to perform operations in the following order: (1) exponents from left to right; (2) multiply and divide from left to right; (3) add and subtract from left to right. Operations inside grouping symbols such as paren-

theses and brackets, as well as inside absolute value notation and radicals are performed before doing them outside grouping symbols, absolute value notation, and radicals. Following is a summary of what simplify means.

Simplify - Order of Operations

The teacher can have students perform operations in the following example in different orders, so that they understand why it is important that mathematicians worldwide agree on the order in which operations are performed. Students will discover that the result varies, depending upon the order the operations are performed. In the example below, the results are 7, $\frac{11}{3}$, and $\frac{2}{3}$.

$20 \div 4 + 6 \div 3$	$20 \div 4 + 6 \div 3$	$20 \div 4 + 6 \div 3$
$5 + 6 \div 3$	$5 + 6 \div 3$	$20 \div 10 \div 3$
$5 + 2$	$11 \div 3$	$2 \div 3$
7	$\frac{11}{3}$	$\frac{2}{3}$

Inside parentheses, brackets, absolute value notation, and radicals.
Perform operations inside parentheses, brackets, absolute value notation, and radicals from inner to outer.

[third (second $\sqrt{\text{first}}$ |first| second) third]

$[\quad 1 - (4 + \sqrt{25-16} - |2-9| + 6) \quad - 5\ ]$ first square root and absolute value

$[\quad 1 - (4 + \quad\sqrt{9}\quad - |\text{-}7| + 6) \quad - 5\ ]$ first square root and absolute value

$[\quad 1 - (4 + \quad 3 \quad - \quad 7 \quad + 6) \quad - 5\ ]$ first square root and absolute value

$[\quad 1 \qquad\qquad\qquad - (6) \qquad\quad - 5\ ]$ second inside parentheses

$\qquad\qquad\qquad -10$ third inside brackets

The order in which square root, absolute value, parentheses, and brackets are performed depends upon their location. The operations located between the innermost notation are done first, followed by the operations in the next innermost notation, and so forth.

Exponents.

An exponent indicates the number of times the base is a factor. Variables can be thought of as dimensions when multiplying and dividing. A foot squared can be written as $(ft^1)^2$. The exponent two means that the base (ft^1) appears as a factor twice $(ft^1 \times ft^1)$. Therefore, $(ft^1)^2 = ft^1 \times ft^1$. A $ft^1 \times ft^1 = ft^2$ since the exponent two indicates the number of times the base is a factor. Teachers could ask students to look for a quicker way to get the result ft^2. Some students will discern that instead of expanding the expression $(ft^1)^2$ to $ft^1 \times ft^1$ and then adding the exponents to get ft^2, they can multiply the exponents in the expression $(ft^1)^2$ and obtain the same result. In summary, the following could be written.

$$(ft^1)^2 = ft^1 \times ft^1 = ft^{1+1} = ft^2 \qquad \text{or} \qquad (ft^1)^2 = ft^{1 \times 2} = ft^2$$

The definition of exponent dictates that exponents are added when multiplying bases. Since division is the inverse of multiplication and subtraction is the inverse of addition, dividing bases means that exponents are subtracted. A cubic ft (ft^3) divided by a foot (ft^1) means that one dimension is separated leaving square feet (ft^2). Therefore, $\dfrac{ft^3}{ft^1} = ft^{3-1} = ft^2$. The following is a summary of the exponent laws.

<u>Exponent Laws</u>
The letter "b" represents a base and the letters "m" and "n" are exponents.

Multiply exponents: $(b^m)^n = b^{mn}$ $\qquad (a^m b^n)^n = a^{mn} b^{mn} \qquad \left(\dfrac{a^m}{b^m}\right)^n = \dfrac{a^{mn}}{b^{mn}}$

Add exponents: $\qquad b^m b^n = b^{m+n}$

Subtract exponents: $\dfrac{b^m}{b^n} = b^{m-n}$

What if m-n is zero or a negative number? Since math is the study of patterns, let's discover what the exponent zero and negative exponents symbolize by looking for a pattern in the following examples.

$5^3 = 125$

$5^2 = 25 \ (125 \div 5)$

$5^1 = 5 \ \ (25 \div 5)$

$3^3 = 27$

$3^2 = 9 \ (27 \div 3)$

$3^1 = 3 \ (9 \div 3)$

$x^3 = (x)(x)(x)$

$x^2 = (x)(x)$ $\quad [(x)(x)(x) \div (x)]$

$x^1 = (x)$ $\quad [(x)(x) \div (x)]$

The teacher could ask students what is done to 125 to get 25 and to 25 to get 5. They will say that each amount is divided by 5, the base. Likewise, the teacher could ask students what is done to 27 to get 9 and to 9 to get 3. They will respond that each amount is divided by 3, the base. Let's see if this pattern also works with variables. The amount $(x)(x)(x)$ divided by (x) is $(x)(x)$ and $(x)(x)$ divided by (x) is x. While the amounts are divided by its respective base, the exponents are decreasing by one. Let's continue this pattern of dividing by the base which decreases the exponent by one to discover what zero and negative exponents mean.

$5^0 = 1 \quad (5 \div 5)$

$3^0 = 1 \quad (3 \div 3)$

$x^0 = 1 \quad (x \div x) \quad x \neq 0$

It appears that any base to the exponent zero is one, that is, any base except zero, $(\text{any base except zero})^0 = 1$. The base cannot be zero because division by zero is impossible. Let's divide 1 by its respective base and see what happens. Remember that the base can be written as base^1; $\text{base} = \text{base}^1$; $5 = 5^1$; $3 = 3^1$; $x = x^1$.

$5^{-1} = \dfrac{1}{5^1} \ (1 \div 5)$

$3^{-1} = \dfrac{1}{3^1} \ (1 \div 3)$

$x^{-1} = \dfrac{1}{x^1} \ (1 \div x) \quad x \neq 0$

The teacher can have students note that the exponent negative one (-1) results in the base written in the denominator to the exponent positive one (1); $5^{-1} = \dfrac{1}{5^1}$; $3^{-1} = \dfrac{1}{3^1}$; $x^{-1} = \dfrac{1}{x^1}$. The students could divide these results by their respective bases to ascertain that $5^{-2} = \dfrac{1}{5^2}$; $3^{-2} = \dfrac{1}{3^2}$; and $x^{-2} = \dfrac{1}{x^2}$.

$$5^{-2} = \frac{1}{5^2} \quad (\frac{1}{5^1} \div 5) \qquad 3^{-2} = \frac{1}{3^2} \quad (\frac{1}{3^1} \div 3) \qquad x^{-2} = \frac{1}{x^2} \quad (\frac{1}{x^1} \div x) \quad x \neq 0$$

A base to a negative exponent in the numerator can be rewritten with the base to the positive exponent in the denominator; $base^{-exp} = \frac{1}{base^{+exp}}$ if base $\neq$ 0. Likewise, the students could discover that a base to a negative exponent in the denominator can be rewritten with the base to the positive exponent in the numerator; $\frac{1}{base^{-exp}} = base^{+exp}$ if base $\neq$ 0.

Following is a summary of the laws for zero and negative exponents.

Any base other than zero divided by itself equals one.

$$(base)^0 = 1 \text{ if base} \neq 0$$

The base cannot be zero because division by zero is impossible.

Negative exponent means to divide by the base raised to the positive of the exponent (base $\neq$ 0).

$$base^{-exp} = \frac{1}{base^{+exp}} \quad \text{and} \quad \frac{1}{base^{-exp}} = \frac{1}{\frac{1}{b^{+exp}}} = base^{+exp} \quad \text{if base} \neq 0$$

$$5^2 = 25 \qquad\qquad 3^2 = 9 \qquad\qquad x^2 = (x)(x)$$

$$5^1 = 5 \quad (25 \div 5) \qquad 3^1 = 3 \quad (9 \div 3) \qquad x^1 = (x) \quad [(x)(x) \div (x)]$$

$$5^0 = 1 \quad (5 \div 5) \qquad 3^0 = 1 \quad (3 \div 3) \qquad x^0 = 1 \quad (x \div x) \qquad x \neq 0$$

$$5^{-1} = \frac{1}{5^1} \quad (1 \div 5) \qquad 3^{-1} = \frac{1}{3^1} \quad (1 \div 3) \qquad x^{-1} = \frac{1}{x^1} \quad (1 \div x) \qquad x \neq 0$$

$$5^{-2} = \frac{1}{5^2} \quad (\frac{1}{5^1} \div 5) \qquad 3^{-2} = \frac{1}{3^2} \quad (\frac{1}{3^1} \div 3) \qquad x^{-2} = \frac{1}{x^2} \quad (\frac{1}{x^1} \div x) \quad x \neq 0$$

The following example illustrates how exponents are performed before multiplying. It also shows students why they do not need to be concerned with negative and zero exponents until after the other exponent laws are applied. Zero and negative exponents often disappear when multiplying, adding, and subtracting exponents.

When exponents are performed first, the result is $7x^6$.

$$7(x^0 x^{-3})^{-2} \qquad \text{add exponents}$$

$$7(x^{-3})^{-2} \qquad \text{multiply exponents}$$

$$7x^6 \qquad \text{no zero or negative exponents remain}$$

If multiplication is performed first, the result is $\frac{1}{49}x^6$.

$$7(1x^0 x^{-3})^{-2} \qquad \text{if multiply coefficients first}$$

$$(7x^0 x^{-3})^{-2} \qquad \text{add exponents}$$

$$(7^1 x^{-3})^{-2} \qquad \text{multiply exponents}$$

$$7^{-2} x^6 \qquad \text{rewrite } 7^{-2}$$

$$\frac{1}{7^2}x^6 \qquad \text{which equals } \frac{1}{49}x^6.$$

3. Multiply and divide from left to right.

When the word *expand* is used as a direction, it refers to the *distributive property* and indicates multiplication. Distribute means to *hand out* which is what distributors of soda do; they dispense pop into machines at schools and businesses. The distributive property hands out a factor to each term in an expression. Expand 4(n+5) means to hand out the 4 to the n and the 5 using multiplication, that is, 4(n+5) is the same as 4 times n plus 4 times 5. In summary, the following could be written.

$$4(n + 5) \qquad \textit{distribute} \text{ the 4 to n and 5}$$
$$4n + 4(5) \qquad \text{multiply 4 times n and 4 times 5}$$
$$4n + 20$$

The following example can be used to illustrate how multiplication and division are performed before addition and subtraction.

$$x^2(x^4) + x^6 \div x^7 \qquad \text{multiply } x^2 \text{ times } x^4$$

$$x^6 + x^6 \div x^7 \qquad \text{rewrite division as a fraction}$$

$$x^6 + \frac{x^6}{x^7} \qquad \text{divide } x^6 \text{ by } x^7$$

$$x^6 + x^{-1} \qquad \text{rewrite } x^{-1} \text{ with a positive exponent}$$

$$x^6 + \frac{1}{x^1} \qquad \text{unlike things cannot be added}$$

4. Add and subtract from left to right.

The order of operations inside parentheses, brackets, or absolute value is the same as when there are no grouping symbols. Exponents are performed first followed by multiplication and division and then addition and subtraction from left to right. Grouping symbols only change the order in which these operations are performed from left to right. The following two examples demonstrate the difference.

$20 \div (2 + 6 \div 3)$	divide inside parentheses		$20 \div 2 + 6 \div 3$	divide first
$20 \div (2 + 2)$	add inside parentheses		$10 + 2$	add
$20 \div 4$	divide		12	
5				

Simplify is certainly a loaded word in mathematics. Students often do not realize all that simplify entails because all its procedures are not usually written and discussed at one time. The teacher could discuss the following outline with students after they have mastered working with grouping symbols and the exponent laws.

Simplify Expressions — Order of Operations

A. Operations inside grouping symbols such as ( ) and [ ], inside absolute value notation, | |, and inside radicals, $\sqrt{}$.

1. Exponents from left to right.

    Multiply exponents: $(b^m)^n = b^{mn}$ $(a^m b^m)^n = a^{mn} b^{mn}$ $\left(\dfrac{a^m}{b^m}\right)^n = \dfrac{a^{mn}}{b^{mn}}$

    Add exponents: $b^m b^n = b^{m+n}$

    Subtract exponents: $\dfrac{b^m}{b^n} = b^{m-n}$

    $(\text{base except zero})^0 = 1$ Any base other than zero divided by itself equals one. Base cannot be zero because division by 0 is impossible.

    $$\text{base}^{-\text{exp}} = \dfrac{1}{\text{base}^{+\text{exp}}} \quad \text{and} \quad \dfrac{1}{\text{base}^{-\text{exp}}} = \dfrac{1}{\dfrac{1}{b^{+\text{exp}}}} = \text{base}^{+\text{exp}} \quad \text{if base} \neq 0$$

    Negative exponent means to divide by the base raised to the positive of the exponent. Base cannot be zero because division by 0 is impossible.

2. Multiply and divide from left to right.
    Expand and the distributive property mean to multiply.
    Variables are dimensions.

3. Add and subtract the coefficients of like terms from left to right.
    Like terms are terms with the same variables.
    Variables are labels.

B. Operations outside grouping symbols.
    1. Same as above.
    2. Same as above
    3. Same as above.

Chapter 17

Pertinent Percent

The following activity is adapted from the book *Make a World of Difference: Creative Activities for Global Learning* . It is a meaningful way to introduce the concept that percent of a total (the original quantity) is an amount. This activity requires that the teacher purchase as many one cent tootsie rolls™ as there are students.

The activity begins with the students calculating the percent of the world population for the five most populated regions: Africa, Asia, Europe, North America, and Latin America which includes Mexico, Central America, and South America. Australia and New Zealand are excluded because they represent less than one percent of the world population. The approximate total world population, excluding Australia and New Zealand, is 5,392,000,000. The following population estimates are from the *1993 Information Please Almanac*.

Region	Population
Africa	654,000,000
Asia	3,491,000,000
Europe	511,000,000
North America	283,000,000
Latin America	453,000,000
Total Population	5,392,000,000

The students calculate and use the percentages listed below to divide themselves into five regions. For the example below, let's assume the class size is 25. The percentage for Europe is actually 9.47% which is rounded to 10% so that the total number of students adds up to 25. In this case, a class of twenty-five students would have three students in Africa, sixteen in Asia, three in Europe, one in North America, and two in Latin America.

Region	World Pop. %	# students
Africa	12%	3
Asia	65%	16
Europe	10%	3
North America	5%	1
Latin America	8%	2
Totals	100%	25

The following percentages represent the distribution of food on these five continents. The teacher distributes twenty-five one cent tootsie rolls® according to these percentages and informs the class that the twenty-five tootsie rolls® represent the amount of food eaten in the world for one day. The number of tootsie rolls® for North America is 5.5 which is rounded to 5 so that the total number of tootsie rolls® is 25.

The three students representing Africa receive 2 tootsie rolls®. The sixteen students in Asia receive only 6. The three students in Europe receive 9. The one student in North America receives 5. The two students in Latin America receive 3. The students might complain that they did not get to choose their continent. The teacher can ask them if people choose where they are born. People do not choose where they are born, but most people live on the continent on which they are born.

Region	World Food %	# tootsie rolls®
Africa	8%	2
Asia	23%	6
Europe	36%	9
North America	22%	5
Latin America	11%	3
Totals	100%	25

It is important to use the same number of tootsie rolls® as the number of students. This allows students to experience the fact that even though there is enough food for every person in the world to eat each day, the food is not distributed this way. Students may want to know if there is anything that they can do to help world hunger. The class could have a jar for money or a box for canned goods whose contents are given to a local food pantry. This activity could be done by all the math teachers in conjunction with an annual school wide food drive.

Teachers could encourage students to do further research on world hunger for their social studies or English class. One organization which can provide information is OXFAM. OXFAM fights hunger by helping impoverished communities achieve self-sufficiency. The address is 26 West Street, Boston, MA, 02111-1206.

The next page has the pertinent information and space for students to do the necessary computations for this activity.

FOOD DISTRIBUTION ACTIVITY

Region	Population
Africa	654,000,000
Asia	3,491,000,000
Europe	511,000,000
North America	283,000,000
Latin America	453,000,000
Total Population	5,392,000,000

Region	World Pop. %	# students	Region	World Food %	#tootsie rolls
Africa			Africa	8%	
Asia			Asia	23%	
Europe			Europe	36%	
N. America			N. America	22%	
Latin America			Latin America	11%	
Total	100%	Class size	Total	100%	Class size

Once students have learned that a percent of a total is an amount, the following activity can be used to demonstrate how percent is used to make predictions.

It is well known that a large percent of the American population drinks alcoholic beverages. Nationally each year, over forty percent of all traffic fatalities are alcohol related. Even though drinking alcohol is illegal for teenagers, a sizable percentage of them drink. Thus, it seems wise that students be taught the mathematics of alcohol consumption and elimination. The following pages contain information for a very worthwhile activity.

The amount of time the body takes to completely eliminate alcohol from a single drink varies according to sex and body weight. The concentration of alcohol is directly proportional to the amount of water in the body. As the weight of a person increases, so does the water content. This varies, however, according to sex and body type. It is estimated that a female's body is approximately 55% water and a male's body is about 68% water. In general, women have less water per pound of body weight than men because women have strategically located fatty tissue which has a lower water content per pound than muscle or bone.

The Swedish scientist, Erik M.P. Widmark, created a formula that estimates the concentration of alcohol in a person's body when drinking on an empty stomach. When drinks are consumed consecutively, the peak alcohol level is usually attained twenty to thirty minutes after the last drink. While drinking on a full stomach lowers the peak level, it slightly lengthens the elimination time.

A single drink is one 12 oz beer, one 4 oz glass of wine, or 1 oz of 100 proof liquor (the proof of liquor equals twice the alcohol concentration; 100 proof is 50% alcohol). Most domestic and import beers are between 4.0% and 4.5% alcohol; lite beers are between 3.3% and 4.1% alcohol; and most dinner wines are between 11% and 13% alcohol.

Male drinking on an empty stomach. Female drinking on an empty stomach.

$$c \approx \frac{a}{9w}$$

$$c \approx \frac{a}{7w}$$

c = concentration of alcohol as a %   beer is between 4.0% and 4.5% alcohol
a = alcohol drank in ml(1oz ≈ 29.6ml)   lite beer is between 3.3% and 4.1%
w = weight in kg (1lb ≈ 2.2kg)   wine is between 11% and 13%

The average rate of elimination is around .015 grams of alcohol per hour on an empty stomach. Because the rate of elimination is constant, the relationship between the time since the first drink and the remaining percentage of alcohol in the blood is linear. When estimating blood-alcohol levels, it is prudent to use the higher alcohol percentage rate for each type of liquor.

$r = -.015t + c$ r = remaining percentage of alcohol in blood
t = time since first drink in hours
c = concentration of alcohol as a percent

Blood-alcohol concentration laws in the State of Wisconsin.

The legal drinking age is 21.

$r = .04$ to $.10\%$ This level can be used as evidence to prove impairment. Reflexes, vision, judgment, and powers of concentration affected. Legal intoxication level for drivers of commercial vehicles.

$r = .10\%$ and above   Legal intoxication level. Physical response seriously impaired.

Students can estimate the concentration of alcohol in a person's blood after consumption of different amounts and types of alcohol. Students can also estimate the number of hours necessary for a person to wait before driving so that the person's blood-alcohol level is below the legal intoxication level of .10%. If students know how to graph linear equations, they can graph the blood-alcohol equation to estimate the time. In order to make these calculations pertinent, students could use a body weight for the person equivalent to their own. It needs to be pointed out that any amount of alcohol in a person under the age of twenty-one is illegal. Below are sample questions and solutions.

Suppose a 140 lb male and a 140 lb female each steadily consume 6 cans of beer on an empty stomach. Use 4.5% as the percent of alcohol in beer to answer the following questions.

1. Estimate the male's and female's concentration of alcohol.

Male drinking on an empty stomach. Female drinking on an empty stomach.

$$c \approx \frac{a}{9w}$$

$$c \approx \frac{a}{7w}$$

c = concentration of alcohol as a %
a = alcohol drank in ml(1oz ≈ 29.6ml)
w = weight in kg (2.2lb ≈ 1kg)

Students must first convert 12 ounces to milliliters and 140 lbs to kg.

$$\frac{1kg}{2.2lb}(140lb) \approx 64kgs$$

$$\frac{29.6ml}{1oz}(12oz) \approx 355ml$$

$$w \approx 64$$

The amount of alcohol(a) in 6 beers must be calculated.
a = (number of beers)(quantity of each beer in ml)(percent of alcohol in beer)
a = (6)(355)(.045)
a = 96ml

The concentration(c) of alcohol can now be calculated.

Male drinking on an empty stomach. Female drinking on an empty stomach.

$$c \approx \frac{a}{9w}$$

$$c \approx \frac{a}{7w}$$

$$c \approx \frac{96}{9(64)}$$

$$c \approx \frac{96}{7(64)}$$

$$c \approx .167$$

$$c \approx .214$$

2. Estimate the time that the male and female need to wait before driving after their first drinks so that their blood-alcohol levels are <u>below</u> the legal intoxication level of .10%. The next lowest reading below .10% on the breath-alcohol measuring device is .09%. Therefore, students need to solve the equation r = -.015t + c when r = .09 or graph the equation and estimate the time.

r = -.015t + c   r = remaining percentage of alcohol in blood
                   t = time since first drink in hours
                   c = concentration of alcohol as a percent

<u>Male drinking on an empty stomach.</u> <u>Female drinking on an empty stomach.</u>

r = -.015t + c	r = -.015t + c
.09 = -.015t + .167	.09 = -.015t + .214
-.077 = -.015t	-.124 = -.015t
5.1 ≈ t	8.3 ≈ t

The male needs to wait at least 5.1 hrs before driving.

The female needs to wait at least 8.3 hrs before driving.

Remember, these calculations are estimates. Therefore, the male would want to wait at least five and a half hours and the female would want to wait at least eight and a half hours before driving. Students discover that a female's blood-alcohol level is much higher than a male's of equal weight when both drink the same amount because women generally have less water in their bodies. Consequently, the elimination time for females is much longer.

The next page has the pertinent information for teachers to design their own activity worksheet. Once students complete the worksheet, teachers can use this activity to facilitate a discussion on the seriousness of drinking and driving. There are alternatives to driving after drinking. Designated drivers can be used and the organization Students Against Drunk Drivers (SADD) offers contracts that students sign with their parents. The contract states that they will call their parents for a ride home, no questions asked, if they find themselves in a situation where they need a ride.

Percent is pertinent to everyone in many ways. This chapter includes just two relevant situations. Students can be encouraged to create and share examples of how they use percent in their lives.

BLOOD-ALCOHOL ACTIVITY

<u>Male drinking on an empty stomach.</u> <u>Female drinking on an empty stomach.</u>

$$c \approx \frac{a}{9w}$$

$$c \approx \frac{a}{7w}$$

c = concentration of alcohol as a %   beer is between 4.0% and 4.5% alcohol
a = alcohol drank in ml(1oz $\approx$ 29.6ml) lite beer is between 3.3% and 4.1%
w = weight in kg (2.2lb $\approx$ 1kg)   wine is between 11% and 13%

The average rate of elimination is around .015 grams of alcohol per hour on an empty stomach. Because the rate of elimination is constant, the relationship between the time since the first drink and the remaining percentage of alcohol in the blood is linear. When estimating blood-alcohol levels, it is prudent to use the higher alcohol percentage rate for each type of liquor.

$r = -.015t + c$ r = remaining percentage of alcohol in blood
                t = time since first drink in hours
                c = concentration of alcohol as a percent

<u>Blood-alcohol concentration laws in the State of Wisconsin.</u>

The legal drinking age is 21.

$r = .04$ to $.10\%$ This level can be used as evidence to prove impairment. Reflexes, vision, judgment, and powers of concentration affected. Legal intoxication level for drivers of commercial vehicles.

$r = .10\%$ and above   Legal intoxication level. Physical response seriously impaired.

Chapter 18

What's Radical about Radicals?

Students are curious about the word "radicals". After all, aren't radicals people who existed in the sixties? Who were some of those people and why were they called radicals? And what does this have to do with math?

Students will mention several names, some of whom might include Rosa Parks, Martin Luther King, Abbie Hoffman, or Malcolm X. The students say that these people were considered radicals because they were trying to change the very nature or basis of society. So a radical is someone who tries to change the basis of society.

Why then is the symbol $\sqrt{}$ called a radical? Let's look at a square root which is one type of radical. <u>Square</u> <u>root</u> actually means the <u>root</u> of the <u>square</u> (exponent 2). It is written $\sqrt[2]{root^2}$ and it equals the root. The root is located beneath the exponent 2, just as the roots of a tree are located at the base. A square root radical can be written as $\sqrt[2]{base^2}$ which equals the base. Simply stated, a radical equals a base.

Just as it is very difficult for radical people to change the basis of society, only certain numbers can be changed to an *exact* base with an exponent. In order to summarize this discussion, the following could be written on the board and copied by students into their journals.

A radical tries to change the <u>basis</u> of society.
A radical in math equals the <u>base</u>.

$\sqrt{}$ is the symbol for radical.
$\sqrt[2]{base^2} = base$

Square root means root of the square.
It is the base below the exponent 2.

$$\sqrt[2]{1} = \sqrt[2]{1^2} = 1$$

$$\sqrt[2]{4} = \sqrt[2]{2^2} = 2$$

$$\sqrt[2]{9} = \sqrt[2]{3^2} = 3$$

$\sqrt[2]{2};\ \sqrt[2]{3};\ \sqrt[2]{5};\ \sqrt[2]{6};\ \sqrt[2]{7};$ and $\sqrt[2]{8}$ have no exact bases.

Cube root means root of the cube. It is the base below the exponent 3.

$$\sqrt[3]{base} = \sqrt[3]{base^3} = base$$

$$\sqrt[3]{1} = \sqrt[3]{1^3} = 1$$

$$\sqrt[3]{8} = \sqrt[3]{2^3} = 2$$

$\sqrt[3]{2};\ \sqrt[3]{3};\ \sqrt[3]{4};\ \sqrt[3]{5};\ \sqrt[3]{6}$ and $\sqrt[3]{7}$ have no exact bases.

Most math books do not include the root 2 outside the radical symbol for square roots. However, students seem to better understand the concept of square root if it is included in the initial presentation. It helps them see that the root and the exponent are equal. Students can be told after the presentation that most math books symbolize $\sqrt[2]{base}$ with $\sqrt{base}$.

Chapter 19

Are Imaginary Numbers for Real?

Mathematicians give names to sets of numbers so that people can use those names to communicate. A set is a group whose members can be listed individually or described in words written between the grouping symbols, { }. The first set of numbers that a child learns is the numbers used for counting, {1,2,3,....}. These are called counting or natural(N) because they are numbers that people naturally use. The three periods at the end of the set symbolize that the pattern continues forever.

When zero is added to this set, the group is called whole(W) numbers, {0,1,2,...}. Students can remember that whole numbers include 0 if they note that the word "whole" contains the word "hole", which looks like a 0. The number zero was added to the number system several years after counting numbers were invented. This is partially because it is difficult to think of or envision *nothing*, which is what zero represents.

Negative numbers are a means of symbolizing left from right, up from down, and gains versus losses. Integers(J) is the set of positive and negative whole numbers, {...,-2, -1 , 0, 1, 2,...}. The capital letter "J" is used to represent the set of integers because capital "I" is used to stand for imaginary numbers which will subsequently be discussed.

All of the above number systems can only describe whole things. Fractions and decimals are used to describe parts of whole things and when they are combined with whole numbers, these numbers are called mixed numbers. Mixed numbers are used to describe whole parts and parts of a whole thing and are called rational(Q) numbers because they can be written as a ratio of two numbers. Students need to be reminded that integers can be written as a ratio whose denominator is one. Rational numbers cannot be listed in a set because there is not a precise fraction immediately preceding or following a number as there is with integers. There always exists another smaller or larger fraction. Therefore, rational numbers are depicted as {fractions or decimals and mixed numbers}. Students can remember that capital "Q" is used to represent rational numbers since a fraction is also a quotient. The letter "R" is not the symbol for rational numbers because it is used to represent real numbers.

Numbers that never end or repeat like π are called <u>ir</u> <u>rational</u>(H) because they cannot be written as a ratio unless they are rounded. <u>Ir</u> <u>rational</u> means <u>not</u> a <u>ratio</u>. Students can remember that the letter "H" is used to represent irrational numbers since these numbers can only be written <u>h</u>orizontally. Other examples of irrational numbers include $\sqrt{2}, \sqrt{3}, \sqrt{5}, \sqrt{6}, \sqrt{7}, \sqrt{8}, \sqrt{10}$, and so forth.

All the above sets of numbers are a subset or part of the set of real numbers. Imaginary numbers are not real numbers. An imaginary number is equal to the square root of a negative number. Since there are no real numbers that equal the square root of a negative number, (two negatives are a positive when multiplying), the Swiss mathematician Henri Euler decided to call them imaginary. For example, $\sqrt{-1}$ equals no real number because (-1)(-1) equals positive one. Instead of always writing the square root of negative one as $\sqrt{-1}$, the letter "i" is used to represent $\sqrt{-1}$.

Euler created the name "imaginary numbers" in the eighteenth century before electricity was discovered and never imagined that these numbers could be used to help solve real problems. Today, electrical engineers use imaginary numbers to design and analyze electrical circuits because imaginary numbers make solving the equations much easier. However, the imaginary parts of the solutions are disregarded. That's right! Imaginary numbers are used to solve real problems but they cannot be a solution because they are not for real.

If all these classifications seem complex, they are! The set of complex numbers includes all numbers, real and imaginary. A complex number is written in the form a+bi, where a and b are real numbers. If a = 0, then the number is imaginary. If b = 0, then the number is real. If both a and b are not 0, the number is called complex because it is composed of both real and imaginary numbers.

In summary, the following list of the sets of numbers in the complex number system could be written on the board and into students' journals. Each set of numbers is contained in the sets that follow directly below it. For example, counting numbers are part of every set except irrational numbers. The set of irrational numbers is written in a column by itself because none of the other sets of real numbers are a subset of it. Irrational numbers are in a class all by themselves. Likewise, the set of imaginary numbers is written in a column by itself because no other set of numbers is a subset of it.

Counting or Natural(N)
{1,2,3,...}
↓
Whole(W)
{0,1,2,...}
↓
Integers(J)
{...-2,-1,0,1,2,...}
↓
Rational(Q)   or   Irrational(H)
{fractions or decimals   {nonrepeating decimals
and mixed numbers}   like π and $\sqrt{2}$}
               ↓   ↓
           Real(R):   Imaginary(I)
   {Rational or Irrational}   { $\sqrt{negative\#}$ }
              ↓   ↓
          Complex Number System(C)
        {Real, Imaginary, or Complex}
Complex numbers are composed of both real and imaginary.

Chapter 20

Multi-Variable Application: Building Stairs

Walking down a staircase, have you ever stumbled on the bottom step because its height is different than that of the other steps? Did you wonder why the bottom step is a different height? It might be because the construction of stairs involves five variables: the number of steps(n), step height(h), step depth(d), the vertical distance of the stairs(v), and the horizontal length of the stairs(L). The following activity can be used to introduce solving simultaneous systems of equations. Students could work in groups of two to design the following stairs.

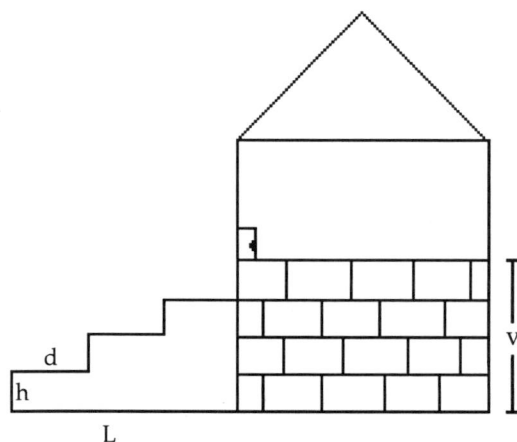

h = height of each step

d = depth of each step

L = length horizontally from the bottom step to the foundation

v = vertical distance from the ground to the threshold

n = number of steps
(The drawing has three steps but the actual number of steps is unknown.)

Building Code
Depth of each step must be at least eleven inches. **d ≥ 11in**

Height of each step must be between four and seven inches inclusive. **4in≤h≤7in**

Let's use the problem solving process discussed in Chapter 12 to design stairs to building code. The code stipulates that the depth of each step must be at least eleven inches (d≥11), and that the height of each step must be between four and seven inches inclusive, $(4 \leq h \leq 7)$.

Step 1: Identify the Problem.

This problem involves writing and solving five equations simultaneously.

Step 2: Develop Specific Strategies.
a. Simplify the problem, if possible.
b. Draw a diagram, construct a table, or plot a graph.
c. Calculate by hand, calculator, or computer.

Since a drawing is provided and the problem cannot be simplified further, the students can begin writing the equations. They could decide to use a spreadsheet to solve the five equations.

Step 3: Translate the Problem into Math. Solve it.

In the picture, the letter "v" represents the distance from the ground to the threshold, the bottom of the door. Since the distance from the ground to the threshold is determined by the construction of the building, this distance is usually known. Let's assume the vertical distance to be 64 inches. Thus, one equation is $v = 64$.

In the drawing, the letter "L" represents the length horizontally from the bottom step to the foundation. In most cases, the distance that stairs extend from a building can vary. For this example, let's see if there are any solutions for height and depth of each step that meets the building code if the horizontal length is 80 inches. Hence, a second equation is $L = 80$.

Three more equations still need to be written. Since the number of steps (n), times the depth of each step (d) equals the horizontal distance (h), a third equation is $nd = L$. Using the same reasoning, students might think that the number of steps times the height of each step equals the vertical distance.

Students need to look closely at the drawing to notice that there is one more height (4) than the number of steps (3). The reason this occurs is because the last step is the threshold, and not another step. Thus, the number of steps plus one (n+1) equals the number of heights (h). In the drawing, the number of steps, three, plus one, equals four, the number of heights. The teacher needs to remind students that the drawing is just a representation of the variables involved in building stairs. The actual number of steps is unknown. A fourth equation can now be written. The number of steps plus one (n+1), times the height of each step (h), equals the vertical distance, $(n+1)h = v$.

The fifth and final equation is the key to constructing stairs with steps of equal height and depth. The number of steps must be chosen so that the step depth and height satisfy the building code (n = ?).

There are several ways to solve these five equations. Students could choose either a depth of 11 inches or more, a height between and/or including 4 and 7 inches, or any number of stairs and then solve the remaining four equations. Students might decide to use a spread sheet to solve this problem when they discover that this system of equations must be solved numerically, not algebraically.

Since the step depth must be at least 11 inches, let's see if the stairs can be designed to code with a depth of eleven inches. The system of five equations is solved when d = 11. A step depth of 11 inches results in $7\frac{3}{11}$ steps. Since the number of steps must be a whole number, students will need to ascertain that the number of steps, "n", must be equal to 7 in order that the step depth be at least 11 inches.

$$v = 64 \qquad L = 80 \qquad (n+1)h = v \qquad \begin{aligned} nd &= L \\ n(11) &= 80 \\ n &= 7\tfrac{3}{11} \end{aligned} \qquad n = ?$$

$$n = 7$$

$$\begin{aligned} 7d &= 80 \\ d &= 11\tfrac{3}{7} \end{aligned}$$

Seven steps results in a step depth of $11\frac{3}{7}$ inches which meets code. Substituting n = 7 into the equation (n+1)h=64 results in a step height of 8 inches which is greater than the maximum code requirement of 7 inches. This situation has no solution that meets code.

$$\begin{aligned} (n+1)h &= 64 \\ (7+1)h &= 64 \\ 8h &= 64 \\ h &= 8 \text{ does not meet code} \end{aligned}$$

In most cases, the horizontal length can vary unless there is a driveway or another building in the way. Let's see if stairs can be built to code if the horizontal length is changed to 110 inches. Substituting the minimum depth of d=11 into the equation nd=110 results in 10 steps.

$$v = 64 \qquad L= 100 \qquad (n+1)h = v \qquad \begin{aligned} nd &= L \\ n(11) &= 110 \\ n &= 10 \end{aligned} \qquad \begin{aligned} n &= ? \\ \\ n &= 10 \end{aligned}$$

Substituting $n = 10$ into the equation $(n+1)h = 60$ results in a step height of $5\frac{9}{11}$ inches which is less than the maximum code requirement of 7 inches.

$$\begin{aligned} (10+1)h &= 64 \\ (11)h &= 64 \\ h &= 5\tfrac{9}{11} \end{aligned}$$

While both of these measurements satisfy the code, measuring tapes are not marked in elevenths. Each inch is divided into halves, fourths, eighths, and sixteenths. Consequently, a builder would need to round nine-elevenths to one of the above measurements, resulting in the height of the bottom step varying from the other steps by a fraction of an inch. As the number of steps increases, rounding fractions can cause the height of the last step to vary by a couple of inches from the other steps.

Some student groups will discover that these stairs can also be built to code with 9 steps of height 6.4 inches and depth $12\frac{2}{9}$. The system of equations is solved when n=9.

$$v = 64 \qquad L = 110 \qquad \begin{aligned} (n+1)h &= v \\ (9+1)h &= 64 \\ 10h &= 64 \\ h &= 6.4 \end{aligned} \qquad \begin{aligned} nd &= L \\ 9d &= 110 \\ d &= 12\tfrac{2}{9} \end{aligned} \qquad n = 9$$

A few student groups will try to say that 8 steps also work. While the depth of 12.75 inches satisfies the building code, notice that the height of $7\frac{1}{9}$ inches is one-ninth of an inch higher than the code requirement. Let's assume that the students decide to construct the stairs with eight steps. The building inspector discovers the discrepancy and tells them to rebuild the stairs to code. Can they afford to build the stairs twice and only get paid once?

$$v = 64 \qquad L = 110$$

$$
\begin{aligned}
(n+1)h &= v & nd &= L & n &= 8\\
(8+1)h &= 64 & 8d &= 110\\
9h &= 64 & d &= 12.75\\
h &= 7\tfrac{1}{9} \text{ does not meet code}
\end{aligned}
$$

Step 4: Check Solution(s) in the Equation(s).

In this case the answers were calculated by substituting them into the equations.

Step 5: Answer Question(s). Choose a Solution.

The owner chooses to have the builder construct 10 steps of approximate height $5\tfrac{9}{11}$ inches and depth 11 inches or 9 steps of height 6.4 inches and approximate depth $12\tfrac{2}{9}$ inches.

Step 6: Check Solution for Practicalness.

Is there room for the stairs to extend from the house 110 inches?

The mathematics involved in designing stairs to building code is an excellent way to show students that math is used to solve practical problems. The students will ascertain that real problems can have more than one possible solution and that some situations have no solution. They will discover that the number of equations must equal the number of variables if a solution(s) is to be found. Students will learn that some systems of equations cannot be solved algebraically.

The next page is an activity worksheet for students to design stairs for $v = 64$ inches and $L = 80$ inches or 110 inches. Students will need to do their work on another piece of paper.

Building Stairs to Code

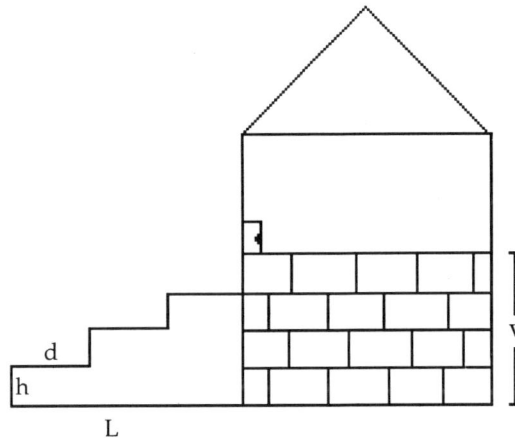

h = height of each step

d = depth of each step

L = length horizontally from the bottom step to the foundation

v = vertical distance from the ground to the threshold

n = number of steps
(The drawing has three steps but the actual number of steps is unknown.)

Building Code

Depth of each step must be at least eleven inches. **d ≥ 11in**

Height of each step must be between four and seven inches inclusive. **4in≤h≤7in**

1. Design stairs for v = 64 inches and L = 80 inches.

2. Design stairs for v = 64 inches and L = 110 inches.

Chapter 21

Slope Encounters of the First Degree

Everyone has encountered slope. It is unavoidable when walking, bicycling, or driving. Slope is both the slant and steepness of a surface. The slant is the direction of the slope, i.e., up, down, horizontal, or vertical. The amount of slope is the steepness.

The slope of a surface going up is considered positive because the direction up is commonly associated with positive. Conversely, the slope of a surface going down is considered negative. Students can draw lines similar to the illustration below in their journals. To discern the difference between a line going up and one going down, they can trace the lines with one of their fingers going left to right. Students trace the lines from left to right because math is read left to right, just as is English.

Some students have difficulty remembering the difference between horizontal and vertical. Since only two options exist, this is a dichotomy. If students remember what one choice is, they will know the other by elimination. The easiest one to remember is horizontal because it contains the root word horizon. Everyone knows the sun "rises" and "sets" on the horizon.

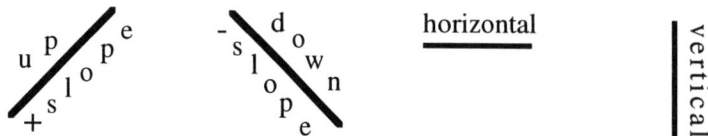

Once students understand the slant of slope, they are ready to learn what steepness is and how it is measured. Mountain roads have warning signs like "steep grade of 6%." Students may know that 6% means 6 out of 100, but they seldom have any idea what 6% has to do with slope.

Since 6% can be written as $\frac{6}{100}$, a steep grade of 6%" means that for every 6 vertical feet the road goes down, it goes 100 feet horizontally. Let's assume this rate of change is constant. The following illustration helps clarify how steepness is measured. Once the picture is drawn, the students will notice that the slope of the road is actually negative six-hundredths, $\frac{-6}{100}$, since the steep grade sign is describing the slant going down the mountain. (The slant would be positive if traveling up the hill.) In summary, the following could be written on the board.

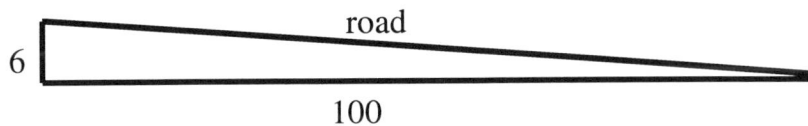

6 road 100

Slope is a constant rate of change. Slope = $\dfrac{\text{constant change in y}}{\text{constant change in x}}$

To find the slope.   1) Determine the slant: +, -, horizontal, or vertical

2) Compute the steepness: $\dfrac{\text{up and down distance}}{\text{left to right distance}}$

Students are now ready to discover that the slope of a horizontal line is zero whereas the slope of a vertical line is undefined. The teacher can have students compute the steepness of two points which lie on a horizontal line. They can see from plotting the two points that the up or down distance is zero. Since zero divided by any number is zero, the slope of a horizontal line is zero. Students can now write "$\dfrac{0 \text{ distance}}{\text{right to left distance}}$ = 0 slope" beneath the horizontal line in the previous illustration.

The teacher can also have students compute the steepness of two points which lie on a vertical line. They can see from graphing the two points that the left to right distance is zero. Because any number divided by zero is impossible, the slope is said to be undefined. Students seem to understand that the slope of a vertical line is "undefined" much better than saying it has "no slope" since they already know that division by zero is impossible. (See the chapter entitled "Division by Zero Gives Meaning to Life.") Some students incorrectly translate "no slope" to "0" because they interpret "no slope"

as "the slope is nothing" which is translated to "0". Students can now write
"$\dfrac{\text{up and down distance}}{0}$ = undefined slope", next to the vertical line in the previous illustration.

This method of finding the slope is easy to remember because a person first thinks of the distance up and down when defining slope, and this is, in fact, what appears first in the numerator. Note that nothing is mentioned about the slope formula. Students are allowed much time and encouraged to discover the formula on their own. If given sufficient time, they eventually get tired of graphing and look for a shortcut. When students discover the slope formula themselves, they remember it as a shortcut to graphing.

In the meantime, students graph two points and draw a line through them. They determine whether the slant is positive (going up), horizontal, negative (going down), or vertical. Then, they compute the steepness: the distance up and down written over the distance left to right.

Chapter 22

Golf Equations

The teacher can begin this lesson by asking students, "How many of you play golf? Those of you who do not play golf, do you know how it is scored?" The students who know golf can explain to the class how it is scored.

A score in golf can be described either by the number of strokes or in relation to par. Most courses have a score of 36 strokes equivalent to 0 par for nine holes. Thus, a score of 37 strokes is equivalent to +1 par and a score of 35 strokes is equivalent to -1 par. The class can discuss why 9 strokes is theoretically the lowest possible score for nine holes. The following chart could be written on the board to help clarify the scoring of golf.

strokes	par
.	.
.	.
.	.
37	+1
36	0
35	-1
.	.
.	.
.	.
9	-27

Since each number of strokes can be paired with a specific par, the paired numbers can be written as ordered pairs. They are called ordered pairs because the order in which the numbers are written matters. The independent variable is written first and the dependent variable is written second.

In order to clarify the difference between independent and dependent variables, a short diversion is necessary. It is helpful to discuss with students why they are claimed as dependents on their parent's tax forms. Students realize that they depend on their parent(s) for food, shelter, and clothing. How much each student depends on their parent(s) for these necessities of life is somewhat determined by their parent(s), the independent variable. In summary, the following could be written on the board.

<u>D</u>ependent variable is <u>d</u>etermined by the independent variable.

The students have enough information now to discover whether strokes or par is the dependent variable. They understand that par is the dependent variable since it is determined by the number of strokes. The following ordered pairs can now be added to the above chart.

strokes	par	ordered pairs
.	.	.
.	.	.
.	.	.
37	+1	(37,+1)
36	0	(36,0)
35	-1	(35,-1)
.	.	.
.	.	.
.	.	.
9	-27	(9,-27)
s	p	(s,p)
ind. var.	dep. var.	(ind.,dep.)

Students could graph several ordered pairs to see the linear relationship between par and strokes. Some students may notice the pattern between strokes and par in the ordered pairs. They see that as strokes increase by 1, the par increases by 1. This *constant rate of change* is, of course, the slope. At this point, students are ready to create a two variable linear equation.

You might be wondering why I am using the words, "two variable linear equation," instead of just "linear equation". I have found that students are frequently confused by the words, "linear equation". For example, $x + 3 = 5$ and $x + 3 = y$ are both linear equations. The first is a one variable linear equation which has only one solution, whereas, the second is a two variable linear equation which has many ordered pair solutions. There are many solutions because these ordered pairs are actually points on a line. Since a line contains an infinite number of points, the equation has an infinite number of solutions.

Several forms can be used to create a two variable linear equation. The slope-intercept form, $y = mx+b$, seems to be the easiest for students to understand. For no known reasons, "m" and "b" are arbitrarily used worldwide as the symbols for slope and the y-intercept.

Let's use the form $p = ms+b$ to create an equation which describes the pattern between the number of strokes and par in golf. Looking back to the ordered pairs in the chart, the two variables are strokes(s) and par(p) It is important to inform students that "s" and "p" are the variables and "m" and "b" are constants. They already know that "m" is a ratio which represents the slope. The meaning of "b" can be explained later when further examples using the form y = mx + b are worked.

Step 1. Find the slope.

Since any two points determine a line, any two ordered pairs can be chosen. Some students might feel comfortable sharing their most recent scores. Let's use (36,0) and (37,1). Assuming that most students do not yet know nor understand the slope formula, the slope can be found by graphing.

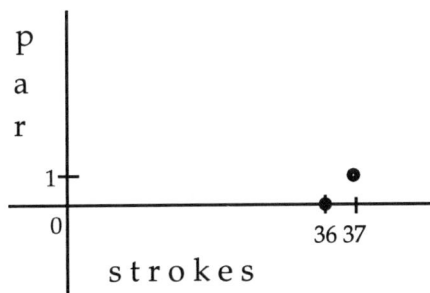

1) Slant is positive, since a line connecting the two points is going up from left to right.

2) Steepness = $\dfrac{\text{up } 1}{\text{right } 1}$

3) Slope = $\dfrac{1}{1}$

Step 2. Substitute the slope for m into p = ms+b.

$$p = \frac{1}{1}s + b$$

Step 3. Find b.
Any ordered pair on the line will make the equation true. Let's substitute (36,0).

$$\begin{aligned}
0 &= 1(36) + b \\
0 &= 36 \quad + b \\
\underline{-36} &\quad \underline{-36} \\
-36 &= \qquad\quad b
\end{aligned}$$

Step 4. Write the equation.

$$p = 1s - 36$$

This is a two variable linear equation which determines the par, given the number of strokes that a person golfs for nine holes. For example, if someone golfs 40 strokes, then that person's score expressed as par is determined to be +4, p = 40-36. If someone else golfs 34 strokes, then that person's par is determined to be -2, p = 34-36.

Since a score of zero strokes is not possible in golf, this pattern does not have a y-intercept. In order to help students understand what the y-intercept, b, represents, teachers could use the following situation.

A school club wants to sell tee shirts as a fund raiser. A local tee shirt company will sell 10 shirts for a $100 or 25 shirts for $205. Let's assume the relationship between cost(c) and the number(n) of shirts purchased is linear. The teacher asks students to write a two variable linear equation to describe this situation.

Step 1. Determine if cost or number of tee shirts is the dependent variable. Since the cost is determined by the number of shirts purchased, cost is the dependent variable.

Step 2. Write two ordered pairs and graph them to find the slope.

(ind.,dep)
(number,cost)
(n,c)

(10,100)
(25,205)

$$m = \frac{up\ 105}{right\ 15}$$

$$m = 7$$

Step 3. Find b. Substitute (10,100) and m = 7 into c = mn + b.

$$c = mn + b$$
$$100 = 7(10) + b$$
$$\underline{-70 \quad -70}$$
$$30 = \qquad b$$

Step 4. Write the equation.

$$c = 7n + 30$$

The teacher asks students to notice that the line crosses or intercepts the y-axis at 30 which is the point (0,30). They can substitute 0 for "n" in c=7n+30 to ascertain that "c" equals 30. Students can discover that "y" is always equal to "b" when "x" is 0 if they substitute 0 for "x" in y = mx + b. The point, (0,b), is called the y-intercept because this is where the line intercepts the y-axis. See the graph below.

Let's discover what the y-intercept represents in this situation. If zero tee shirts are sold, the business figures its expense is still thirty dollars. This original charge goes to defray what are called fixed costs. Fixed costs are expenses that a business has, even if it does not sell any goods. No matter how many tee shirts are imprinted, the business must pay costs of utilities, rent, insurance, etc.

When students learn how to graph horizontal and vertical linear equations, the following information can be added to their notes. Students will better understand and remember that a horizontal line is described by a number equal to "y" if the horizontal axis is also depicted as the line "y=0", since all y's are zeros on the x-axis. Likewise, the vertical axis could be renamed as the line "x=0", since all x's on the y-axis are zeros.

The following summary could be written on the board and entered into the students' journals.

y-intercepts(0,b)

x=0 on y-axis

y=0 on x-axis

<u>To write an equation of the line in the form y = mx + b.</u>

1. Calculate the slope, "m", the constant rate of change.

$$m = \frac{\text{constant change in y}}{\text{constant change in x}}$$

2. Substitute the value for "m" in the equation y = mx + b.

3. Find the y-intercept, "b", which is the point (0,b), located on the y-axis.
   a. Substitute a point on the line for "x" and "y" in the equation $y = mx + b$
   b. Solve for "b".

(When "0" is substituted for "x" in the equation $y = mx + b$, the equation becomes $y = m(0) + b$, which simplifies to $y = b$. Since the y-intercept occurs when x equals zero or the beginning point, it represents the original amount or fixed cost in a situation.)

4. Substitute the values for "m" and "b" in the equation $y = mx + b$.

Golf scores and business fixed costs are just two examples of how two variable linear equations can be meaningfully introduced. These realistic examples give students something to associate with when they work problems from their books that seem meaningless.

Chapter 23

Functions and Chromosomes

This chapter contains much information about teaching the concept of a function. The material needs to be staged over a period of at least three class meetings, because too much information can be presented at one time, impeding students' learning.

The first class session could be used to introduce domain and range. This lesson can begin with a quick review of dependent and independent variables using golf, as was done when teaching two variable linear equations. (See the chapter entitled "Golf Equations.") At this time, the following sentence and chart could be written on the board.

Dependent variable is determined by the independent variable.

strokes	par	ordered pairs
.	.	.
.	.	.
.	.	.
37	+1	(37,+1)
36	0	(36,0)
35	-1	(35,-1)
.	.	.
.	.	.
.	.	.
9	-27	(9,-27)
s	p	(s,p)
ind. var.	dep. var.	(ind.,dep.)

Students typically recall why par is the dependent variable, since it is determined by the number of strokes. The teacher can have them notice, though, that the number of strokes possible differs from the pars that are possible. For nine holes, the possible number of strokes goes from 9 to infinity; the possible pars go from -27 to infinity. Thus, the possible values for the independent variable can be different from the possible values for the dependent variable. Consequently, the two sets of possibilities need different names. The set of possible values for the independent variable is called the domain.

The set of possible values for the dependent variable is called the range. In summary, the following can be written on the board.

> Domain: Set of possible values for the independent variable.
> Domain of golf for nine holes: {9,10,11, . . .}
>
> Range: Set of possible values for the dependent variable.
> Range of golf for nine holes: {-27,-26,-25, . . .}

All of the above is discussed the day prior to the introduction of a function. The day that a function is introduced, teachers can begin by writing the chart below on the board. Students who were present the day before need only add the last two lines to their charts. Those who were absent can copy the entire chart as the teacher quickly reviews independent and dependent variables, domain and range.

strokes	par	ordered pairs
.	.	.
.	.	.
.	.	.
37	+1	(37,+1)
36	0	(36,0)
35	-1	(35,-1)
.	.	.
.	.	.
.	.	.
9	-27	(9,-27)
s	p	(s,p)
ind. var.	dep. var.	(ind.,dep.)
domain	range	(domain,range)
x	y	(x,y)

The letters "x" and "y" are included in this chart to remind students that in equations of the form $y = mx+b$, "x" is the independent variable and "y" is the dependent variable. Students may recall that the relationship between the number of strokes and par was determined to be linear when two variable linear equations were taught. The teacher can quickly recreate the equation.

Step 1. Calculate the slope, "m".

Let's assume most students have discovered the slope formula by the time this lesson is taught. Let's use (37,1) and (36,0) as the two points (x_1, y_1) and (x_2, y_2).

$$m = \frac{y_2 - y_1}{x_2 - x_1} \qquad m = \frac{1-0}{37-36} \qquad m = \frac{1}{1}$$

Step 2. Substitute the value for "m" into $y = mx + b$.

$$y = \frac{1}{1}x + b$$

Step 3. Find b, the y-intercept.

Any ordered pair from the chart will make the equation true. Let's substitute (36,0).

$$
\begin{aligned}
0 &= 1(36) + b \\
0 &= 36 + b \\
\underline{-36} &\quad \underline{-36} \\
-36 &= \qquad b
\end{aligned}
$$

Step 4. Write the equation.

$$y = 1x - 36$$

Since "x" represents strokes(s) and "y" stands for par(p), the equation can be rewritten using the letters "s" and "p".

$$p = 1s - 36$$
$$p(s) = 1s - 36$$

That's correct. Both p and p(s) represent par. The notation p(s) is read "p of s" which means p is determined by s. If someone golfs forty strokes, then that person's par is *determined* to be four, $p(40) = 40 - 36$, or $p(40) = 4$. The teacher can have students notice that the ordered pair (40,4) is contained in the equation $p(40) = 4$.

$$p(40) = 40 - 36$$
$$p(40) = 4$$
$$(40,4)$$

The notation p(s) is used to describe a function, which is the topic of this lesson. In order to help students understand what a function is, the teacher can ask them once again to refer to the chart of ordered pairs on the board. The teacher can have them notice that each number of strokes is paired with only one par. Thirty-seven strokes is paired with only plus one par; thirty-six strokes is paired with only zero par; and so forth. In other words, each "x" is paired with only one "y". Not many students understand exactly what this means. There is an association, however, that most students will understand and remember.

Some students may recall from biology that the "x" chromosome determines a female and the "y" chromosome determines a male. (Students can remember that the "x" chromosome represents female(f), and "y" chromosome stands for male(m) by recognizing that "x" comes before "y" just as "f" comes before "m" in the alphabet.) They are also well aware of the double standard that applies to females and males in relationships. It is more accepted in our society for males to have more than one female relationship at a time than it is for females to have more than one male relationship at a time. While this is the case, teachers need to make it very clear that they are not condoning such a double standard. The purpose for discussing these relationships is that a function describes a relationship with the same double standard. That is, each "x" (female) is paired with only one "y" (male), but that each "y" (male) can be paired with more than one "x" (female).

After a lively discussion, the following summary can be written on the board and in students' journals.

Functions have a double standard.

A function is a relationship whereby each "x" (female) is paired with only one "y" (male), but each y (male) can be paired with more than one x (female).

In order to ascertain if students fully understand the above description of a function, students could classify the ordered pairs listed below as either a function or not a function. Assume x_1, x_2, x_3 represent three different girls and y_1, y_2, y_3 stand for three boys. Students can explain their responses in writing.

(x_1, y_1)	(x_1, y_1)
(x_1, y_2)	(x_2, y_1)
(x_1, y_3)	(x_3, y_1)
not a function	a function

The students still do not know why a function has restrictions on "x". It certainly is not because of some double standard that exists in life. Let's look at why the above situation on the left is not a function. Students can graph the subscripts as ordered pairs, i.e., (1,1) and (1,2) and (1,3) to see what happens.

They will see that the line connecting the points is a vertical line whose slope is $\frac{1}{0}$ which is undefined. Each "x" can be paired with only one "y" because, when the same "x" is used more than once, the line connecting these points is a vertical line whose slope is undefined. And functions do not want to deal with undefined situations. Students can now summarize in their journals why a function restricts each element of the domain to only one member of the range.

Ordinarily, it is not wise to explain "why" and "how" a concept works, simultaneously. In this case, however, the students have previously learned "why" division by zero is impossible. (See the chapter entitled "Division by Zero Gives Meaning to Life.") The second day of staging a function is now complete.

On the third day, the concept of mapping is presented. The following examples include variables since most math books have variables in mapping exercises.

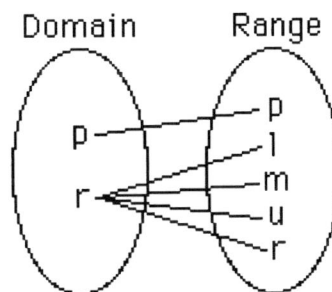

The above drawing means that p is paired with or mapped to p, and r is paired with or mapped to l, m, u, and r. Thus the ordered pairs (p,p); (r,l); (r,m); (r,u); and (r,r) describe the relationship between the domain and the range. Since the domain member "r" has more than one range member, this relationship is not a function.

Students will want to know what kind of relationship these letters could possibly represent. At this time, the following can be written on the board.

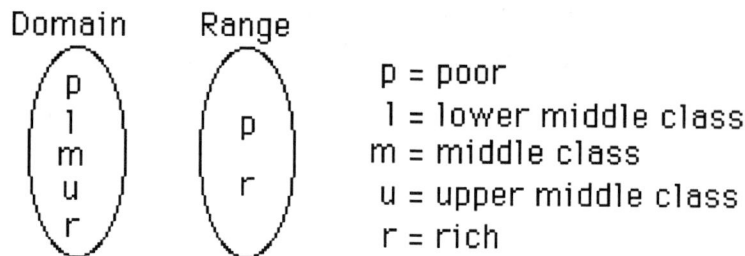

```
Domain     Range
   p
   l
   m            p
   u
   r            r
```

p = poor
l = lower middle class
m = middle class
u = upper middle class
r = rich

Let's assume that the domain represents the economic class structure of people living in a developed country like USA, and that the range represents the economic class structure of people living in an undeveloped country like Nicaragua. Many students are aware that undeveloped countries generally have only poor and rich people, no middle classes. After a short discussion on what the above information represents, teachers can ask students to map the economic class of people in the USA to the economic class of people in Nicaragua. Most students draw their arrows as in the following drawing.

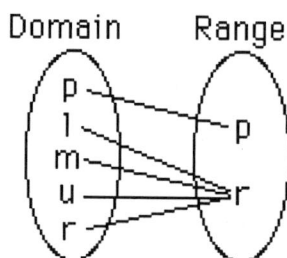

```
Domain     Range
   p
   l
   m            p
   u
   r            r
```

No matter how students draw their arrows, this exercise shows them that variables can and do have meaning. In this case, a variable is mapped to another variable.

The class could discuss other relationships where a variable is mapped to a number. For example, a menu could be described by mapping each item to its price. The domain would be the items on the menu; the range would be the prices. This discussion concludes the third day of staging a function.

Chapter 24

Liability Step Functions

Liability car insurance rates provide an excellent way to introduce step and discontinuous functions. Students can construct graphs comparing liability rates to a person's age, depending upon their sex. Following are the annual liability amounts for males and females who are owners of a single car. The rates are lower for students who maintain at least a B average in school. These rates apply as long as they remain a full-time student. If they graduate from college these rates apply until they are twenty-five years old.

Male Annual Liability Rates

16 to 21	$778
21 to 25	$460
25 to 30	$260
30 and above	$230

Male "B average" Rates

16 to 21	$595
21 to 25	$355
25 to 30	$260 (no discount)
30 and above	$230 (no discount)

Female Annual Liability Rates

16 to 21	$422
21 to 25	$336
25 and above	$230

Female "B average" Rates

16 to 21	$403
21 to 25	$317
25 and above	$230 (no discount)

The teacher could discuss with the class how statistics are used to determine these rates. Males between the ages of 16 and 21 pay the highest rate and are rewarded the biggest amount of savings for maintaining at least a B average. Statistics show that young males with grade averages below a B have the highest rate of accidents.

After students have constructed four separate graphs illustrating the information in the above charts, they can draw a graph comparing male to female rates. Below is the graph comparing males' liability rates to their ages.

Annual Liability Costs for Males

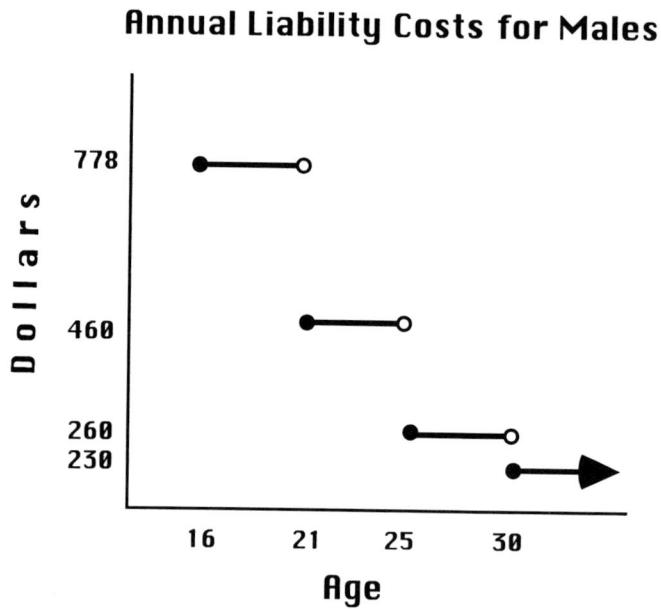

This application illustrates why step functions are discontinuous. The graph jumps to another level because the liability rate skips from one amount to another, depending upon the person's age and sex. This application helps students understand why there are open and closed circles above each other in the greatest integer function graph. It also explains why a graph would have an arrow attached to the last step.

Chapter 25

Polynomial Functions, Equations, and Expressions

Students can more fully understand a concept when it is compared and contrasted to other related concepts. This chapter compares and contrasts linear, quadratic, and cubic polynomial functions, equations, and expressions.

Whenever possible, it is important to relate math concepts to another subject because it helps students who have a good understanding in that other subject. Most students know the difference between a phrase and a sentence. If they do not, then they will learn it in math class.

A phrase is an incomplete thought; it makes no sense unless it is part of a sentence. Likewise, an expression makes no sense unless it is part of an equation. An equation is a sentence with a subject, predicate, and a predicate noun. Equations do not have direct or indirect objects since the equal sign can be read "is", which is a being verb. Thus, one side of the equation renames (predicate noun) the other (subject). This comparison of expressions and equations to phrases and sentences helps students, who know sentence structure, understand math.

Equations are problems to be solved whereas expressions are merely phrases that can be rewritten or simplified. Simplifying procedures are skills required to solve equations. This is comparable to skills needed in music to play an instrument or in sports to play a game. Functions are special equations which show the relationship between at least two variables. Functions are graphed so that projections can be made by seeing the relationship between the variables. Following is a summary of definitions that apply to polynomial functions, equations, and expressions.

Equation:	A sentence of two expressions joined by an equal sign.
Expression:	Phrase of terms.
Term:	Factors separated by a + or - sign.
Factors:	Numbers and variables that make products.
Polynomial:	Many terms of the form ax^n where $n \geq 0$ and $a \in R$.
Degree:	Highest sum of variable exponents within a term.
Linear:	Degree equals one.
Quadratic:	Degree equals two.
Cubic:	Degree equals three.

Charts are helpful in visualizing comparisons. Together, the teacher and students can construct a chart similar to the one at the end of the chapter. The teacher begins by asking for examples of linear, quadratic, and cubic expressions, equations, and functions. The chart contains examples of each. The teacher asks students if they see a difference between equations and functions. Some students will notice that all polynomial functions are equal to "y". The students can graph the functions to observe why this is true. A polynomial function can always be set equal to "y" because this guarantees that each "x" will be paired with only one "y".

The words that give the directions for functions, equations, and expressions are also included in the chart. Expressions are simplified, expanded, factored, or evaluated. Equations are solved, or the roots or solutions are found. Functions are graphed and the x-intercepts are the zeros of the function. The zeros of a function are found by setting the function equal to zero, that is, "y" is zero. Since an x-intercept means that "y" is zero in an ordered pair (x,0), finding the zeros of a function means locating the x-intercepts.

In order to help students visualize the above information, they could draw the following graph in their journals. The y-axis is renamed the "y intercepts (0,y)" and the x-axis is renamed the "x-intercepts (x,0)" or "zeros of a function".

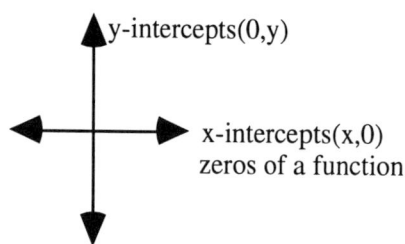

y-intercepts(0,y)

x-intercepts(x,0)
zeros of a function

The following chart is beneficial to students since almost everything that is studied in high school math courses is either an expression, equation, or function. It puts into perspective how they are interrelated. For example, expressions have no equal sign whereas equations and functions do. Since an equal sign means balanced, any operation can be performed to both sides of an equation or function without disturbing its equilibrium. Because expressions have no equal sign, they can only be multiplied or divided by one since anything "times" or "divided by" one does not alter the original expression. This comparison helps students understand when to multiply by the common denominator (equations) versus when to get a common denominator (expressions).

Polynomial Functions, Equations, and Expressions

Expressions	Equations	Functions
linear: x	x = 1	x = y
linear: x + 3	x + 3 = 4	x - 1 = y
quadratic: x^2+2x-1	$x^2+2x-1=2$	$x^2+2x-3=y$
cubic: x^3-5x^2+2x+9	$x^3-5x^2+2x+9=1$	$x^3-5x^2+2x+8=y$
phrase	sentence	sentence subject is y
No equal sign	Has equal sign	Always equals y
Simplify	Solve	Graph
Expand or Factor	Find root(s)	Find zero(s)
Evaluate	Find solution(s)	Find x-intercept(s)
Multiply or Divide by one	Perform an operation to both sides of equation.	Perform an operation to both sides of function.
Skills	Games or Performances	Special Relationship

Chapter 26

Solving Polynomial Equations

As mentioned in the last chapter, an equation is two expressions joined by an equal sign. The same operation can be performed to both sides of an equation without altering the solution because equal means balance. If two people balance a teeter totter everyone knows what happens when one person jumps off. It certainly does not remain balanced! However, if two people of the same weight sit on opposite sides of a balanced teeter totter, it remains level. Following is a summary of the strategies used to solve linear, quadratic, and higher order polynomial equations.

Method for Solving One Variable <u>Linear</u> Equations.

    A. Overall strategy is to isolate the variable.

        If equation has fractions or decimals, multiply both sides by LCD.

        1. Simplify any expressions, if possible.

        2. Add inverses to both sides.

        3. Multiply both sides by the reciprocal of the coefficient.

Method for Solving One Variable <u>Quadratic</u> Equations.

    A. Overall strategy is to set the equation equal to zero.

        1. Factor, if possible.

            a. Factor out the common factor.

            b. Use FOIL factoring.

        2. Use the quadratic formula.

Method for Solving One Variable <u>Higher Order</u> Equations with one or more rational solutions.

    A. Strategy is to use synthetic division until it is reduced to a quadratic.

        1. Factor, if possible.

            a. Factor out the common factor.

            b. Use FOIL factoring.

        2. Use the quadratic formula.

This brief summary gives students a perspective on how the strategies used to solve polynomial equations of any degree are related. Students could write the above in their journals along with examples that illustrate they know when and how each strategy is applied.

Chapter 27

Factors are like Factories!

Do your students have difficulty remembering the difference between factors and multiples? Years ago, a girl in my pre-algebra class created a mnemonic device for factors. A mnemonic is simply a technique for remembering information and is not to be used to replace understanding of the material. If used properly, a mnemonic can also enhance a person's understanding of the information, as her device does.

During a class discussion of factors, she stated that factors are like factories. The class laughed, myself included, since we assumed that she was joking. She insisted, however, that she was serious and that she wanted me to write the following on the board.

Factories: <u>machines</u> <u>make</u> <u>things</u>
Factors: <u>numbers</u>   <u>make</u> <u>the answer when multiplying</u>

The class quickly saw the similarity between factors and factories. Both make something. In factories, machines make things. In math, numbers and variables make the answer when multiplying.

As I was writing the words "the answer when multiplying," I realized the beauty of what this student had created! I asked the class if anyone knew what "the answer when multiplying" is called. Someone responded "product". Then I asked if anyone knew what "the things made in factories" are called. Again someone said "products". I now wrote the following on the board. The entire class was impressed and agreed that they had no excuse for not remembering what factors are.

Factories: <u>machines</u>   <u>make</u> a <u>product</u>
Factors: <u>numbers and variables</u>   <u>make</u> a <u>product</u>

In the expression xy, the factors are x and y because xy is the same as x times y. The factors of 6 are 1 and 6 since 1x6 = 6 and 2 and 3 because 2x3 = 6. The number 6 is the product and called a multiple of 1, 2 ,3, and 6.

Students might want to create a mnemonic device for multiple. This is unnecessary. Comparing factors to multiples represents a dichotomy, i.e., in a problem such as 2x3 = 6, a number is either a factor or a multiple. In a dichotomy there are only two possibilities, if one of them is known (what factors are); the other one can be determined by elimination. If it is known that factors are *numbers and variables that make* a product, then the *product* (the other possibility) must be the multiple.

Factoring is a very important skill in math because it changes an addition and/or subtraction expression into a multiplication expression. Finding the solution to a multiplication problem is easy if the equation is set equal to zero, since any amount of zeros is zero: (any amount)(0) = 0. The following example illustrates one important use of factoring.

<u>Factors can be used to solve quadratic equations.</u>

$n^2 - 3n + 1 = 5$	solution unknown
$\quad\quad -\underline{5} \;\; -\underline{5}$	subtract 5 from both sides
$n^2 - 3n - 4 \;\; = 0$	equation is equal to zero
$(n+1)(n-4) = \;\; 0$	factor

If either of the factors (n+1) and (n-4) is zero, then the equation will be solved since any amount times zero equals zero. The number that makes each factor zero is found by setting each factor equal to zero and then solving for the variable.

$$n + 1 = 0 \quad \text{and} \quad n - 4 = 0 \qquad \text{set factors equal to zero and solve}$$
$$\underline{-1} \;\; \underline{-1} \underline{+4} \;\; \underline{+4}$$
$$n = -1 n = \; 4$$

Since -1 and 4 make the factors zero, they are the solutions to this equation. The solutions, -1 and 4, can be substituted into the equation $n^2 - 3n+1 = 5$ to ascertain their validity.

Factors are also used to simplify rational, decimal, and fraction equations. The terms of the equation are factored to find the lowest common denominator (LCD). Since equal means balanced, each term of the equation can be multiplied by the LCD without altering the solution(s).

Factors can simplify Rational Equations by multiplying both sides by the GCF.

$$\frac{3}{4n+20} = \frac{n}{n+5}$$

$$\frac{3}{4(n+5)} = \frac{n}{n+5}$$

$$(n+5)\frac{3}{4(n+5)} = \frac{n}{n+5}(n+5)$$

$$\frac{3}{4} = n$$

Factors can simplify Decimal Equations by multiplying by the LCD.

$$.3x + .38 = 1.58$$

$$(100).3x + .38(100) = 1.58(100)$$

$$\begin{array}{rcl} 30x + 38 & = & 158 \\ -38 & & -38 \end{array}$$

$$30x = 120$$

$$x = 4$$

Factors can simplify Fraction Equations by multiplying by the LCD.

$$\frac{2x}{3} + \frac{4}{5} = 6$$

$$(15)\frac{2x}{3} + \frac{4}{5}(15) = 6(15)$$

$$10x + 12 = 90$$

$$10x = 78$$

$$x = 7.8$$

Factors are also used to reduce fractions and rational expressions. It is often difficult to ascertain what expression of one is to be divided by without factoring the numerator and denominator. The expression 3n+15 can be written as the factors 3(n+5) since 3 is a common factor of 3 and 15. Likewise, the expression 4n+20 can be written as the factors 4(n+5) because 4 is a common factor of 4 and 20. Consequently, the fraction $\frac{3n+15}{4n+20}$ can be rewritten as $\frac{3(n+5)}{4(n+5)}$. Since the numerator and denominator are written as multiplication expressions, this fraction can be easily divided by one, expressed as $\frac{(n+5)}{(n+5)}$.

<u>Factors can be used to reduce Rational Expressions.</u>

$$\frac{3n+15}{4n+20} = \frac{3(n+5)}{4(n+5)}$$

$$\frac{3(n+5)}{4(n+5)} \div \frac{(n+5)}{(n+5)} = \frac{3}{4}$$

<u>Factors can be used to reduce Fractions.</u>

$$\frac{6}{8} = \frac{3}{4}\left(\frac{2}{2}\right)$$

$$\frac{3}{4}\left(\frac{2}{2}\right) \div \left(\frac{2}{2}\right) = \frac{3}{4}$$

In summary, students can define factors and explain their importance. Factors are the variables and numbers that make up the product. When the word factor is used as a direction it means to write the given addition and/or subtraction expression as a multiplication expression. Factors are important because they can be used to reduce fractions and rational expressions; to simplify rational, decimal, and fraction equations; and to solve quadratic equations. (In more advanced math courses, it could be stated that factors can be used to solve polynomial equations of degree two or more.)

Chapter 28

Progressions: Aging of Cats and the Power of Love

Cats do most of their growing in the first year of their life. Consequently, veterinarians estimate that kittens grow to adulthood(human age 18) in one year. Each year thereafter, they estimate that one cat year is equivalent to six human years based upon the average life span of a cat. Of course, this is just an educated guess. When cats are asked their age, they enigmatically respond "meow."

With students' input, the teacher constructs the following table.

human years	18	24	30	36	......	h
cat years	1	2	3	4	......	c

In small groups, the students generate the formula below which describes the relationship between human and cat years. The teacher may need to help some groups get started formulating this equation.

$$\text{human years} = 18 + 6(\text{cat years} - 1)$$

Since there is a common difference of six years between the human and cat years, the relationship between human and cat years forms an arithmetic sequence. A sequence is a progression of numbers such as 18, 24, 30, 36, etc. The numbers are called the terms of the sequence and symbolized by t_n. Each term is numbered(n) based on its position in the sequence starting with one.

In this example, the cat years(c) "number each term" in the human years sequence. When c = 1, the first term(t_1) is 18. When c = 2, the second term(t_2) is 24 and so forth. The teacher now appends a row to the above table entitled "term number" so that the chart looks like the following.

human years	18	24	30	36	......	h
cat years	1	2	3	4	......	c
term number	1st	2nd	3rd	4th	......	n

$$\text{human years} = 18 + 6(\text{cat years} - 1)$$
$$h = 18 + 6(c - 1)$$

18 is the <u>first</u> <u>t</u>erm, represent it with t_1

6 is the common <u>d</u>ifference, symbolize it with <u>d</u>

c is the term's <u>n</u>umbered position in the sequence, use <u>n</u>

h is the <u>n</u>th term, express it as t_n

The formula for generating a cat's age into human years can be general-ized into the following equation which describes arithmetic sequences.

<u>Arithmetic Sequences:</u> <u>A progression of numbers with a common difference.</u>

$$h = 18 + 6(c - 1)$$
$$t_n = t_1 + d(n - 1)$$

t_1 = <u>first</u> <u>t</u>erm

d = common <u>d</u>ifference

n = term's <u>n</u>umbered position in the sequence ($n \in N$)
1st, 2nd, 3rd, 4th, etc.

t_n = <u>n</u>th <u>t</u>erm

Once students have the above information copied into their journals, the class will want to compute and discuss the equivalent human age of their cats. This discussion gives students the opportunity to experience math as a language that they can use to model reality. Students enjoy this lesson!

Geometric sequences can also be introduced by modeling reality. Let's assume a person has been treated rudely by another person. There is a high probability that the rude person's behavior will elicit a negative response from the offended person, who in turn will likely offend someone else, and so on. How can this circle of negativity be broken?

One person may not be able to change the world, but s/he can certainly effect her/his immediate world, e.g., family, friends, and community by positive acts of behavior. Let's assume that one person decides to treat two others in a positive manner. Let's also assume that each person who experiences a positive encounter will treat two others in a positive manner. At the first level of encounters, one person positively affects two people. At the second level of encounters, both of these two people positively influence two others for a total of four people. At the third level of encounters, each of these four people treat two others in a positive manner for a total of eight people, and so on. In order to clarify this situation, the following assumption and sketch could be written on the board.

Assumption: Each person who experiences a positive encounter will treat two others in a positive manner, and these two people will each treat two others in a positive manner, and so on.

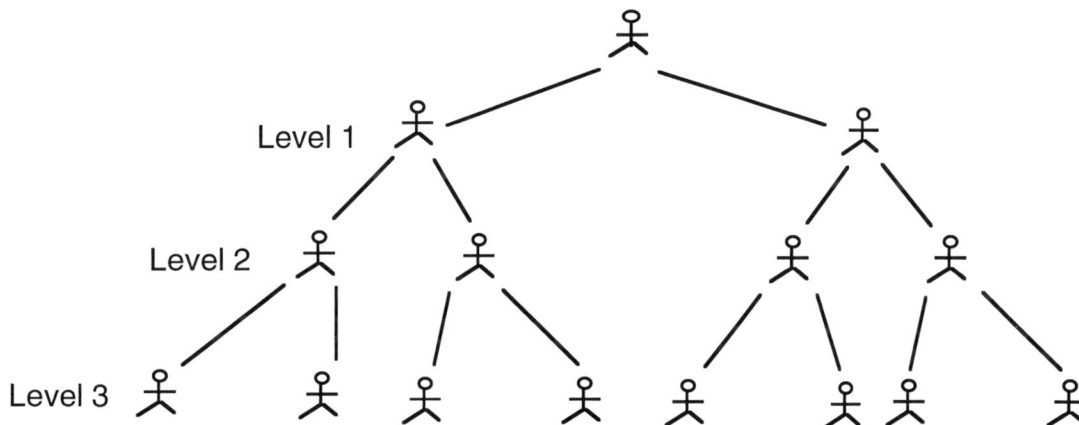

With the students' input, the teacher constructs the following table on the board.

level number	1	2	3	4	......	n
affected people	2	4	8	16	......	p

In small groups, the students generate the following formula which describes the relationship between the number (n) of encounters and the quantity of people (p) affected.

$$p = 2^n$$

The teacher can show students how $p = 2(2)^{n-1}$ is equivalent to $p=2^n$ by adding the exponents [1+(n-1) is n]. The equation is rewritten in this form in order to explain how geometric progressions are generated. The level number (n) is the term's numbered position in the sequence. The affected people (p) is the nth term (t_n).

$$p = 2(2)^{n-1}$$

The first 2 is the first term (t_1).

The second 2 is the common ratio (r).

n is the term's numbered position in the sequence (n)

p is the nth term (t_n)

The above formula can be generalized into the following equation which describes geometric sequences.

Geometric Sequences: A progression of numbers with a common ratio.

$$p = 2(2)^{n-1}$$

$$t_n = t_1 r^{n-1}$$

t_1 = first term

r = common ratio

n = term's numbered position in the sequence ($n \in N$)

$$t_n = n^{th} \text{ term}$$

Once students have the above information copied into their journals, the class could discuss other examples that progress geometrically. Some students might explain that the above formula could also describe how the AIDS virus can be transmitted, if we assume that each person has only two sexual partners. As far as the virus is concerned, each person is having sex with the partner's two previous partners, and with the two previous partners' partners, and so on, to the umpteenth power of two!

Using mathematical modeling to introduce arithmetic and geometric sequences, shows students that there is personal value in learning mathematics. Math can be used to make sense of the world.

Chapter 29

The Value of Absolute Value

The value of goods and services is their monetary worth in the marketplace, which of course is a positive amount. Imagine an item being sold for a negative amount, that is, the buyer is paid for purchasing the item! Absolute means a principle or rule is true in all cases. The value of absolute value is always non-negative (zero or positive). Since distance is never negative, absolute value can be thought of as the distance from zero.

$\left|0\right|$ = 0, because 0 is zero spaces from zero.

$\left|1\right|$ = 1, because 1 is one space from zero.

$\left|-1\right|$ = 1, because -1 is one space from zero.

Absolute value equations and inequalities need to be staged over three class periods, since students have great difficulty understanding this concept. After experimenting with several explanations, I discovered an approach that most students seem to understand. Each problem is translated into English and simplified to one of three forms. Then a number line is drawn.

On the first day, students are shown how to translate the three basic forms of absolute value problems into English. Once written in words, students can solve the problem by understanding what they wrote or by drawing a number line to see the answer. Students copy the following three cases into their journals. Let's assume the domain is the set of real numbers. The word "spaces" is used instead of "units" because students seem to better understand "spaces".

Case 1. $\left|x\right| = 2$ Numbers two spaces from zero.

Of course, 2 and - 2 are two spaces from 0.

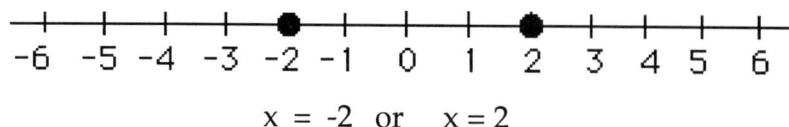

$$x = -2 \quad \text{or} \quad x = 2$$

Case 2. $|x| < 2$ Numbers less than two spaces from zero.

Since the numbers are located between -2 and 2, "x" is written between -2 and 2 (-2 < x < 2).

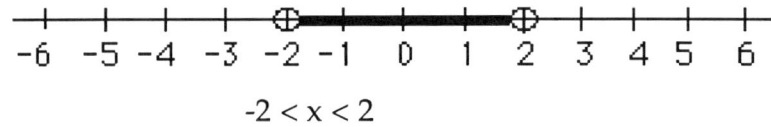

-2 < x < 2

Case 3. $|x| > 2$ Numbers more than two spaces from zero.

The numbers more than two spaces from zero are to the left of -2 (less than -2) or to the right of 2 (greater than 2).

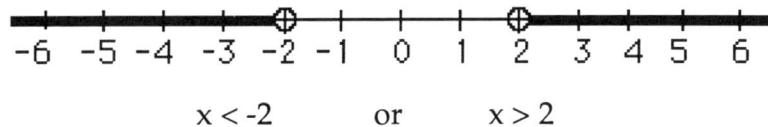

x < -2   or   x > 2

Translating this problem into words explains why there are two answers and why the inequality sign changes from greater than (x > 2) to less than (x < -2). The sign changes because numbers located to the left of a number are less than that number.

All absolute value equations and inequalities can be simplified to one of these three forms. On the second day of staging this concept, students learn how to simplify more complex problems by substituting a variable for a more complex expression. Students copy the following three examples into their journals.

Ex.1 $|2x + 3| = 5$ Let u = 2x + 3

$|u| = 5$ Substitute u for 2x +3. Equation has been simplified to one like Case 1.

u = 5   or   u = -5   Numbers 5 spaces from zero.

2x+3 = 5   or   2x+3 = -5
2x   = 2   or   2x   = -8
 x   = 1   or   x   = -4

Ex.2 $\left|2x - 5\right| < 3$ Let u = 2x - 5

        $\left|u\right| < 3$ Substitute u for 2x - 5. Equation has been
                                                      simplified to one like Case 2.

        $-3 < u < 3$ Numbers less than 3 spaces from zero.

        $-3 < 2x - 5 < 3$
        $\underline{+5 \qquad +5 \quad +5}$
        $\;\;2 < 2x \;\;\; < 8$
        $\;\;1 < \;\;x \;\;\; < 4$

Ex.3 $\left|4x - 6\right| > 2$ Let u = 4x - 6

        $\left|u\right| > 2$ Substitute u for 4x - 6. Equation has been
                                                      simplified to one like Case 3.

        $u < -2$ or $u > 2$ Numbers more than 2 spaces from zero.

        $4x-6 < -2$ or $4x-6 > 2$
        $4x \;\; < 4$ or $4x \;\;\; > 8$
        $\;\;x \;\; < 1$ or $\;\;x \;\;\; > 2$

   This completes the second day of staging absolute value. The third day
is spent discussing the following four exceptions.

Exception 1. $\left|x\right| = 0$ Numbers zero spaces from zero.

          Of course, 0 is the only number zero spaces from 0.

          $\left|x\right| = 0$
          Solution: x = 0.

Exception 2. $\left|x\right| = -2$ Numbers a negative two spaces from zero.

          Since distance is non-negative, there are no numbers at
          negative two spaces from zero. The solution of the abso-
          lute value equal to any negative number is always the
          empty set.

          $\left|x\right|$ = any negative number.
          Solution: x = ø.

Exception 3. $|x| < -2$ Numbers less than negative two spaces from zero.

> Since the numbers less than negative two are negative and distance is always non-negative, there is no solution. The solution to the absolute value less than any negative number is always the empty set.

> $|x| <$ any negative number.
> Solution: $x = \emptyset$.

Exception 4. $|x| > -2$ Numbers greater than negative two spaces from zero.

> Since all numbers are located a non-negative space from zero, all numbers in the domain are greater than negative two spaces from zero. The solution to the absolute value greater than any negative number is the domain.

> $|x| >$ any negative number.
> Solution: $x =$ domain.

Students will be able to solve absolute value equations and inequalities if they translate them into words. If students simply follow the patterns discovered in cases one through three, without translating them into English, they will get incorrect solutions. Once problems are written in words, students find the solutions by comprehending what they wrote and by drawing number lines to see the answers.

Chapter 30

Lincoln Logs: Common Logarithms and the Power of ABE

A logarithm is a synonym for exponent. The word "log" means exponent. Logarithmic form is just another way of expressing exponential form and is useful in describing and graphing scientific applications. The Richter scale, which is used to measure the magnitude of earthquakes, is exponential or logarithmic. Every increase of one unit on the Richter scale represents a growth in magnitude of an earthquake by one power of ten (10^1). An increase of two on the scale symbolizes a growth of magnitude to the second power of ten (10^2). For example, an earthquake of magnitude seven is 10 times (10^1) greater than a magnitude six earthquake and 100 times (10^2) greater than a magnitude five earthquake.

In order to help students see the similarity between exponential and logarithmic forms, the following could be written on the board and into their journals.

Exponent equals log	log equals Exponent
E = log	log = E

Exponential Form	Logarithmic Form
$A = B^E$ if B>0	$\log_B A = E$ if B>0

A = Amount	<u>A</u>mount is written next to log.
B = Base	<u>B</u>ase is written <u>B</u>elow log.
E = Exponent	<u>E</u>xponent is a <u>log</u>arithm.

Students easily perceive that the letters spell ABE and are the first letters of words that define its meaning. Since a logarithm is an exponent, "log" must be set equal to "E", the exponent (log = E). Mathematicians decided to position the amount, <u>A</u>, next to log and place the base, <u>B</u>, below and to the right of log ($\log_B A$). Since mathematicians are always looking for shortcuts, the base is assumed to be 10 when no base is written. Students need to be informed that capital letters are required in the logarithmic equation because small "e" represents a number. The following two cases could now be worked.

Case 1. Write $64 = x^3$ in log form.

$$A = B^E$$

$$\log_B A = E$$

$$\log_x 64 = 3$$

Case 2. Write $\log_5 y = 3$ in exponential form.

$$\log_B A = E$$

$$A = B^E$$

$$y = 5^3$$

If students write the two original equations using ABE, as done in the above two examples, they can then actually see what A, B, and E represent. The teacher needs to have students ascertain that the amount will always be greater than zero (A>0) since the base is always greater than zero (B>0). Once students master how to translate from the exponential to logarithmic form and vice versa, they are ready to learn the product, quotient, and power theorems of logarithms. The logarithmic theorems can be introduced by comparing and contrasting them to the exponent rules. Students need to think of "logA" as an "exponent".

1. Power Theorem: Multiply Exponents.

$$(\log A)^x = (x)(\log A)$$

$$(\log 10^2)^3 = (3)(\log 10^2)$$

$$\log 10^6 = (3)(2)$$

$$6 = 6$$

An exponent raised to an exponent means to multiply exponents.
Both logA and x are exponents.

2. <u>Product Theorem:</u> <u>Add Exponents.</u>

$$\log A_1 A_2 \quad = \quad \log A_1 + \log A_2$$

$$\log 10^3 10^4 \quad = \quad \log 10^3 + \log 10^4$$

$$\log 10^7 \quad = \quad 3 + 4$$

$$7 \quad = \quad 7$$

When multiplying amounts, the exponents are added.

3. <u>Quotient Theorem:</u> <u>Subtract Exponents.</u>

$$\log \frac{A_1}{A_2} \quad = \quad \log A_1 - \log A_2$$

$$\log \frac{10^3}{10^4} \quad = \quad \log 10^3 - \log 10^4$$

$$\log 10^{-1} \quad = \quad 3 - 4$$

$$-1 \quad = \quad -1$$

When dividing amounts, the exponents are subtracted.

The teacher might want to work an example which illustrates that

$$\log \frac{A_1}{A_2} \neq \frac{\log A_1}{\log A_2}.$$

$$\log \frac{100}{10} \neq \frac{\log 100}{\log 10}$$

$$\log 10 \neq \frac{2}{1}$$

$$1 \neq 2$$

In the expression $\log \dfrac{A_1}{A_2}$, the amounts can be divided first, then the log calculated, if both amounts are numerical. For example, $\log \dfrac{100}{10}$ can be rewritten as $\log 10$ which equals one. Likewise, the amounts can be multiplied first, if they are both known, as in $\log(100)(10)$ $[\log(100)(10) = \log 1000 = 3]$.

The letters "ABE" can also be used to introduce the concept of anti-log. "Anti" is a prefix which means opposite. Since the base is given, the opposite of the exponent is the amount. The <u>a</u>ntilog of the exponent is the <u>a</u>mount(<u>A</u>). The <u>a</u>ntilog$_B$E = <u>A</u> is also translated into $A = B^E$ in exponential form.

<u>A</u>ntilog$_B$E = <u>A</u> is written in exponential form as $A = B^E$.

This lesson is an intuitive approach of comparing exponent laws to the symbolism of logarithmic theorems. After students are comfortable with the symbolism, the teacher could show them how to prove the logarithmic theorems using the laws of exponents.

Chapter 31

Pi e: Irrational Parameters of Life

Most students know that π is approximately equal to 3.14, but few understand what π represents. The following activity allows students to discover the meaning of π by measuring the circumference and diameter of two cans. It can be used in any math course from middle school through precalculus. This activity can also be used to teach the concept of a radian.

Materials needed for pi activity.
Ruler for each student.
An empty pop can for half of the students in a class.
An empty small juice can for half of the students in a class.
A 25 cm piece of string for each student.

In order to effectively manage this activity, it is a good idea to have the needed materials on students' desks when they arrive in class. The teacher distributes the juice and pop cans so that students sitting next to each other have a different type of can. The definitions of circumference and diameter, along with the following chart, is written on the board.

Circumference(c): The <u>distance</u> around a circle.
Diameter(d): The <u>distance</u> of a chord passing through the center of a circle.

Can Type	Circumference(c)	Diameter(d)	$\frac{\text{circumference}}{\text{diameter}}$ $(\frac{c}{d})$
Juice			
Pop			

Students first copy the definitions and chart from the board into their journals. They then measure the circumference and diameter of the top of the juice or pop can to the nearest millimeter. If given time, students will determine that the ruler can be used to estimate the diameter but that both the string and ruler are necessary to approximate the circumference. While half the class is measuring the circumference and diameter of the juice can, the other half is measuring the pop can. When students finish measuring the can distributed to them, they exchange cans with their neighbor. After students have taken measurements of both cans, they compute the quotients of the circumferences divided by their diameters. Most students realize that their results are very close to π. The following summary could be written on the board and into students' journals.

$$\frac{c}{d} \approx 3.14$$

$$(d)\frac{c}{d} \approx 3.14(d)$$

$$c \approx 3.14d$$

But what does c ≈ 3.14d mean? In order to help students discover the meaning, they remeasure, with the string, the circumference of either can. Next, students use the string length of the circumference to measure the number of diameters in the circumference. They discover that approximately 3 diameters go around the circumference of a can. Students add the following to their notes.

c   ≈   3.14d

circumference is approximately three diameters

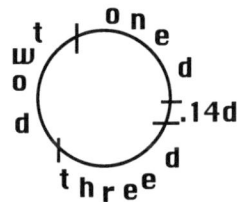

If this activity is being used to introduce the concept of a radian, the teacher could continue with the following discussion. Since two radii equal the diameter, the number of radii around the circumference is twice the number of diameters in the circumference; c ≈ 3.14d can be rewritten as c ≈ 3.14(2r) or c ≈ 6.28r.

$$c \approx 3.14d$$

$$c \approx 3.14(2r)$$

$$c \approx 6.28r$$

$$c \approx 6.28 \text{ radii}$$
$$c \approx 6.28 \text{ radians}$$

A radian is equivalent to the radius. A radian is the arc length of a circle equal to the length of the radius. The following equivalent ratios can be written to show how one radian is approximately equivalent to fifty-seven degrees.

$$\frac{6.28 \text{ radians}}{360 \text{ degrees}} = \frac{3.14 \text{ radians}}{180 \text{ degrees}} \approx \frac{1 \text{ radian}}{57 \text{ degrees}}$$

$$\frac{2\pi \text{ radians}}{360 \text{ degrees}} = \frac{\pi \text{ radians}}{180 \text{ degrees}} \approx \frac{1 \text{ radian}}{57 \text{degrees}}$$

Mathematicians worldwide agree to use the Greek letter pi(π) to represent the number of the diameters around the circumference. Students can remember the number pi is associated with a circle because it is pronounced pie. Pi is called irrational because this number cannot be expressed as a ratio unless it is rounded.

This activity is worthwhile since students tend to understand a concept better and retain it longer if they are actively involved in the learning process. As a result of this activity, many students will be able to recall the meaning of π.

The letter "e" is also used to represent an irrational number. Mathematicians agree to use "e" because it is the first letter of the surname <u>E</u>uler (pronounced Oiler). Euler was an eighteenth century Swiss mathematician who proved the existence of this irrational number. Dependent upon the students' level of knowledge, they can discover that "e" is approximately 2.72 by computing the sum of the first six terms in the infinite series $1 + \frac{1}{1!} + \frac{1}{2!} + \frac{1}{3!} + \frac{1}{4!} + \frac{1}{5!} + ... + \frac{1}{n!}$ or they can use their calculator to estimate "e" as the $\lim_{x \to \infty}(1 + \frac{1}{x})^x$.

When the irrational number "e" is used as a base of an exponential or logarithmic function, it can model real life applications. The logarithm to base e is called the *natural* logarithm because it can be used to model the continuous decay or growth of *natural* occurrences such as the growth of yeast and bacteria or the decay of radioactive material. Students are totally amazed that a number that never ends or repeats could be used to describe patterns in nature. The number "e" can also be used to describe the continuous decrease and increase of other phenomena such as the growth of money invested at an interest rate compounded continuously.

Math textbooks contain numerous applications using the continuous growth and decay equation, $f = be^{rt}$. This formula, however, differs slightly from textbook formulae in that the letters "b" and "f" are used to represent the beginning and final quantities, respectively. If this formula is being used to generate a quantity that does not grow or decay *continuously*, the formula produces only an approximation.

$$f = be^{rt}$$ where

b = beginning quantity
f = final quantity
e ≈ 2.72
t = time
r = rate of continuous decay (r<0) or
   rate of continuous growth (r>0)

Pi and "e" are two irrational numbers that are used to rationally explain naturally occurring phenomena.

y = cars: y = c + asinr(x-s)

Most math books write the above equation using letters that have no association with what they represent. In the above equation the letters are the first letters of words that describe what they symbolize. In groups of two, students can explore what these letters represent. The discovery of each letter's representation is a separate activity. It is best that no more than two activities occur during one class period. Students can graph each equation by hand or on a graphics calculator. Of course, a graphics calculator allows the students to explore many more graphs.

Activity I. Students explore the letter "**c**" in y = c + sinx.

Students graph y = sinx, y = 2 + sinx, and y = -2 + sinx in their journals, even if they are using a graphics calculator. The teacher allows each group to discover that "c" represents the center of each graph. That is, half of the curve is above the center line and half is below it, therefore the center line is the average of the extreme values of "y." It is important to mention that the graph moves up or down because "c" is being added to sinx which is equal to "y," and "y" is graphed up and down. In summary, the following could be written into students' notebooks beneath their graphs.

y = c + sinx

c = center line

Activity II. Students explore the letter "**a**" in y = asinx.

Students graph y = sinx, y = 2 sinx, and y = .5sinx in their journals. The teacher allows time for each group to see that "a" represents the maximum distance of the curve above and below the center line. Students who have studied wave theory in physics will detect that "a" stands for amplitude. Most students will need an explanation of what amplitude measures. Amplitude is the maximum distance above or below the center line. The teacher needs to point out that the amplitude, by definition, is a distance, therefore, it is always positive. A negative amplitude does not exist.

Students could now graph y = -sinx, y = -2sinx, and y = -.5sinx to investigate how a negative sign before "a" effects the graph by comparing them to the original three graphs. They will perceive that the sign before the letter "a" just signifies in which direction the curve starts. Students can note that negative sine graphs *start* in the downward direction. Students will discover, when they inspect y = c + acosr(x-s), that a negative sign preceding "a" causes the graph to *face* downward. In summary, the following could be written into students' journals under their graphs.

$$y \ = \ \mathbf{a}sinx$$

$$\left| \mathbf{a} \right| \ = \ amplitude$$

Amplitude is the <u>maximum distance</u> above or below the centerline.

Activity III. Students explore the letter "**r**" in y = sinr(x).

Since some students will snicker at the mention of the word period, it is wise for the teacher to begin this activity by discussing the physiological meaning of period. Biologically speaking, a period is the interval of time it takes to complete one menstruation cycle. In a similar manner, mathematicians use period to refer to the interval it takes to graph one complete cycle of a curve.

Students graph y = sinx from 0 to 720 degrees. They can see that the interval it takes to graph one complete cycle is 360 degrees, the period. Since the period is 360 degrees, students need to observe what is happening in a 360 degree interval in each of the following graphs. If graphics calculators are used, students need to set the domain to 360 degrees.

Students now graph y = sinx, y = sin2x, and y = sin.5x from 0 to 360 degrees in their journals. Allow each group to discover that "r" represents the number of times the cycle repeats in a 360 degree interval. Have students create a formula that relates the number of repeats(r) in the period(p) of 360 degrees. Most students will ascertain that the period equals 360 degrees, divided by the number of repeats ($period = \dfrac{360°}{repeats}$). Some students may recognize that, in physics, "r" is called the frequency of a wave, which means the number of times a wave repeats in one period. In summary, the following could be written into students' journals beneath their graphs.

$$y = \text{sin}r(x) \qquad r = \text{repeats} \qquad period = \frac{360°}{repeats} \qquad p = \frac{360°}{r}$$

Period is the interval it takes to graph one complete cycle of a wave.
Repeat means the number of times the sine wave repeats in one 360° period.

Activity IV. Students explore the letter "**s**" in y = sin(x - **s**).

Students know from activity III that the period of a sine curve is 360 degrees. In order for students to see what "s" does to the graph, the domain for the following graphs could be from -360 to 720 degrees. Students can graph y = sinx, $y = \sin(x - 90°)$, and $y = \sin(x + 90°)$ in their journals.

The teacher allows each group to discover that "s" represents the number of degrees the graph is shifted to the left or right of the origin. The graph shifts left or right of the origin because "s" is subtracted from "x" in y = sin(x - s) and "x" is graphed left to right. The teacher can have students notice that the graph $y = \sin(x - 90°)$ shifts to the right. Substituting 90° into $(x - 90°)$ results in 0, which shifts the graph 90 degrees to the right of the origin. The graph $y = \sin(x + 90°)$ shifts to the left. Substituting $-90°$ into $(x + 90°)$ results in 0, which shifts the graph 90 degrees to the left.

At the end of this fourth activity, the following summary could be written on the board and into the students' journals.

$$y = c + asinr(x - s)$$

```
      e    m    e    h
      n    p    p    i
      t    l    e    f
      e    i    a    t
      r    t    t
      l    u
      i    d
      n    e
      e
```

<u>C</u>enterline is the average of the extreme values of "y".

<u>A</u>mplitude is the <u>maximum</u> <u>distance</u> above or below the centerline.

<u>R</u>epeats mean the number of times the wave repeats in one period.

<u>S</u>hift is distance the graph shifts to the right or left of the origin.

Period is the interval it takes to graph one complete cycle of a wave.

$$period = \frac{360°}{repeats} \qquad\qquad p = \frac{360°}{r}$$

Students can use the above four activities to explore the translations of the other trigonometric functions. These activities can also be used to write an equation to describe trigonometric graphs. When writing an equation to depict a graph, students can observe the period from the graph and solve for "r" in $p = \frac{360°}{r}$.

Teachers can refer to the equation y = c + sin(x-s) when they explain translations of non-trigonometric functions. Students understand how to graph equations of the form $y = c + |x\text{-}s|$ or $y = c + \sqrt{x\text{-}s}$ after a quick comparison to what the letters "c" and "s" represent in y = c + sin(x-s).

Chapter 33

Geometric Proofs and the Law

The process used to prove deductive geometric proofs can be introduced by comparing and contrasting it to the procedure lawyers use to verify a case in court. In law, there are certain strategies that are accepted as evidence in proving a case. Lawyers use definitions, laws, and precedents to substantiate their cases beyond a reasonable doubt. Mathematicians use definitions, properties, postulates, axioms, and previously proven theorems to prove other theorems.

Definitions are necessary so that communication can be effective. Laws are rules of conduct that society has agreed to live by in order to establish and maintain a civilized world. Laws are based upon common societal beliefs and morals, but they cannot be proven right or wrong. Stealing cannot be proven wrong. Nonetheless, it is against the law to steal because nobody likes to be "ripped off". In mathematics, properties, postulates, and axioms are laws which cannot be proven. There is a postulate in geometry that states that any two points determine a line. It cannot be proven that any two points determine a line, yet everyone knows this from experience, and therefore accepts it as a law.

Precedents are previous rulings on a case of law. They are very important because they can be used to prove future similar cases. In mathematics, theorems are like precedents in that they are specific cases, which once proven, can be used to prove other cases or theorems.

In law, sometimes there are pictures relevant to the case. In 1991, Los Angeles police officers were videotaped beating Rodney King. The defense and prosecuting attorneys deduced different facts from observing the same video. In order to avoid different interpretations of geometric pictures, mathematicians have agreed that nothing can be assumed to be true from merely observing a figure. Given facts will either be marked with appropriate, accepted symbols in a figure, or placed in writing.

Both lawyers and mathematicians prove cases by first identifying what they are attempting to prove and then using definitions, laws and precedents to substantiate their cases. The following outline could be written on the board to help clarify the comparison of the process lawyers and mathematicians use to prove cases.

I. Proving a case.

   A. Identify what is to be proven.

   B. Prove the case.

      1. Lawyers use the following to verify a case.

         a. Definitions.

         b. Laws — cannot be proven right or wrong.

         c. Precedents — previous rulings on similar cases.

      2. Mathematicians use the following to prove a case.

         a. Definitions.

         b. Laws — called postulates, axioms, and properties.

         c. Precedents — previous proven cases called theorems.

Chapter 34

The Fundamental Principle of Rigidity

The following activities will help students discover the two dimensional figure that is the most rigid and why it is the most rigid. They will also learn that not just any three lengths can be formed into a triangle. This activity starts with an extra credit assignment that is due within one week. Students are to construct a triangle and another non-triangular plane figure out of any material, e.g., wood, or straws and pipe cleaners.

All figures are displayed in the classroom so that students can physically test the shapes to feel which is the most rigid. After all students have had an opportunity to do this, the teacher can ask them what they discovered. Most of them will say that a triangle is the most rigid. In order to help them learn why a triangle is the most rigid plane figure, the following activity can be performed with five different lengths of rope: one, five, seven, nine, and twelve feet.

Before students learn why a triangle is the most rigid, the ropes can help students ascertain that not any three lengths can form a triangle. Three students can try to use the one, five, and nine foot ropes to form a triangle on the floor. Students conjecture in their journals why a triangle cannot be formed. Another three students can use the five, seven, and twelve foot ropes to attempt to form a triangle. Once again, students conjecture why a triangle cannot be formed. In order to summarize what they have discovered, students could write the following in their journals.

> The sum of any two lengths must be greater than the third length in order to connect them to form a triangle.

Next, the seven, nine, and twelve foot ropes are formed into a triangle with the teacher and two students standing on the ends of the ropes. The teacher asks the class what the three people standing represent. They recognize that the people depict points. The teacher now takes the two rope ends that s/he is standing on and places them on her/his desk. The two students may need to move closer to the desk in order to make this possible. The teacher stands on both rope ends on the desk. That's right! The teacher stands on her/his desk and asks the class why the ropes still form a triangle even though s/he is not standing on the floor. Someone will respond that the three standing people are still in a common plane. That is, three points are in the same plane no matter where the points are positioned.

The teacher asks another student to stand on her/his chair. The rope connecting the two students on the floor, is used along with a fourth rope, to form a four sided figure. The class will see that the figure is not planar. The teacher can have students notice, though, that any three of the standing people are in a common plane.

Why is a triangle the most rigid figure? Because any three non- collinear points are in the same plane. This concept was known to the builders of the pyramids and it is a fundamental principle of construction today. The teacher can have students give specific examples of structures where triangles are visible.

In order to summarize this activity, students could write the following in their journals.

A triangle is the most rigid figure because any three non-collinear points are in the same plane.

Chapter 35

Counting on Your Fingers is not Immoral

Counting on your fingers is not immoral, but mathematics can be used for immoral purposes. Years ago I heard a speech given by John Gofman, a highly respected scientist who worked several years for the Atomic Energy Commission. He was also a member of the Manhattan Project group which produced the first atomic bomb. He sadly stated that he wished one of his math or science teachers would have informed him that applications of mathematics are not value-free. That is, mathematics is used somewhere in the process of manufacturing every product, which can be classified as to its value to the world.

Imagine that each product on the market has a life enhancement/endangerment rating. Every commodity is labeled using a scale from 1 to 10. One represents a rating that is 100% life enhancing and ten represents 100% life endangering. Such a rating system might get people to think about whether they want or need to purchase a given commodity. Perhaps people would realize that they can live without some things that are life threatening or purchase an alternate item that is more life enriching or environmentally sound, e.g., glass instead of plastic bottles.

To help students realize that math applications are not value-free, the teacher can divide the class into groups of three. Each group lists and rates three products: one that is soundly environmentally safe, one whose life enhancement outweighs its endangerment, and one that is definitely harmful to the planet. Once the groups have generated their examples, they share their information with the entire class. Each student justifies how a rating was assigned to the product and explains how math was used to produce it. Students who describe an item that is definitely harmful to the environment, suggest an alternate product that is environmentally safer.

Students enjoy this cooperative learning activity because it gives them an opportunity to apply the value of math to life. The process of assigning a life-value to products is not easy, but it makes students aware of an object's usefulness for life. And, as in life, there is no one correct answer.

References and Recommended Readings

Adapted from *Experiencing More with Less*, Meredith Sommers Dregni, Herald Press, 1983 in *Make a World of Difference: Creative Activities for Global Learning.* Friendship Press, 1989.

Glasser, William M.D. *The Quality School: Managing Students Without Coercion.* Harper & Row, 1990.

National Council of Teachers of Mathematics. *Curriculum and Evaluation Standards for School Mathematics.* NCTM, 1989.

National Council of Teachers of Mathematics. *Professional Standards for Teaching Mathematics.* NCTM, 1991.

National Fire Protection Association. *Life Safety Code Handbook.* 1988.

National Research Council. Banchoff, Thomas F. *On The Shoulders of Giants: New Approaches to Numeracy.* National Academy Press, 1990.

National Research Council. *Everybody Counts: A Report to the Nation on the Future of Mathematics Education.* National Academy Press, 1989.

Paulos, John Allen. *Innumeracy: Mathematical Illiteracy and its Consequences.* Farrar, Strats & Giroux, 1988.

Paulos, John Allen. *Beyond Numeracy.* Random House, 1992.

Polya, G. *How To Solve It.* Princeton University Press, 1988.

Simon, Sidney B. et al. *Values Clarification:: A Handbook of Practical Strategies for Teachers and Students.* Dodd, Mead, & Company, 1985.

Watzlawick, Paul. *The Language of Change.* Basic Books, 1978.

Willoughby, Stephen S. *Mathematics Education for a Changing World.* Association for Supervision and Curriculum Development, 1990.

The Information Please Almanac. Houghton Mifflin, 1993.

INDEX

Order Form

Making Math Matter
by John Mudore
Christa McAuliffe Award Winner

An essential resource for experienced and beginning teachers.
The book lies flat when open to make its use easier.

Who is this book for?
Middle and high school math teachers.
College and university instructors of math courses below calculus.
Professors and students of math methods courses.

What does this book contain?
Classroom management techniques designed specifically to make teaching
and learning math more enjoyable.
Specific teaching strategies on how teachers can make math matter by presenting algebraic, geometric, and trigonometric topics in a real world setting.

Why select this math resource book?
Written by a veteran classroom teacher.
Tested in actual classrooms.
Easily integrated into any curriculum or textbook.
Helps teachers implement the NCTM standards.

Name_____

School_____

Street Address_____

City/State/Zip_____

Number of copies _____ @ $19.95.   Total payment enclosed _____.

Price includes postage and handling.
Make check payable to Infinity Publishers.

Mail check and order to:
Infinity Publishers
P.O. Box 333
Black Earth, WI 53515
(608) 767-2381

About the Author

A veteran classroom teacher, John Mudore was awarded a 1992 Christa McAuliffe Fellowship. The prize is given by the U.S. Department of Education to recognize outstanding teachers and enable them to develop innovative programs.

The author has taught mathematics in a variety of settings: rural and urban areas, poor and affluent districts, private and public schools, traditional and alternative programs. He has been a math instructor at the grade school, junior high, high school, college, and university levels, as well as at a maximum security prison.

For several years, Mudore has shared his expertise in teacher-training workshops. Teachers who have attended his workshops have asked him for written material on his management and teaching strategies. *Making Math Matter* is his response to these requests.

For information about Mudore's teacher-training workshops, contact:

Infinity Publishers
P.O. Box 333
Black Earth, WI, 53515
(608) 767-2381